Galois Cohomology of Elliptic Curves

J. Coates
R. Sujatha

Published for the

Tata Institute of Fundamental Research

Narosa Publishing House

New Delhi Chennai Mumbai Calcutta London

International distribution by

American Mathematical Society

J. Coates
University of Cambridge
Cambridge, U.K.

R. Sujatha
Tata Institute of Fundamental Research
Mumbai, India

N A R O S A   P U B L I S H I N G   H O U S E

6 Community Centre, Panchsheel Park, New Delhi 110 017
22 Daryaganj, Prakash Deep, Delhi Medical Association Road, New Delhi 110 002
35–36 Greams Road, Thousand Lights, Chennai 600 006
306 Shiv Centre, D.B.C. Sector 17, K.U. Bazar P.O., New Mumbai 400 705
2F–2G Shivam Chambers, 53 Syed Amir Ali Avenue, Calcutta 700 019
3 Henrietta Street, Covent Garden, London WC2E 8LU, UK

ISBN 81-7319-293-6

Published by N.K. Mehra for Narosa Publishing House, 6 Community Centre,
Panchsheel Park, New Delhi 110 017 and printed at Replika Press, Pvt Ltd,
Narela Delhi 110 040, (India).

Galois Cohomology of Elliptic Curves

*Sokidaku mo
Ogironaki kamo
Kokibaku mo
Yutakeki kamo
Koko mireba
Ube shi kamiyo yu
Hajimekerashi mo

What a sweep of sea,
The vastness of these waters!
What abundant life
In these teeming, boundless deeps!
One has but to look —
No wonder from the age of gods
Began the building of this site.

*From the *Man'yōshū*. See *A Waka Anthology, Vol. 1: The Gem-Glistening Cup* by E.A. Cranston, Stanford University Press, Stanford, 1993.

Preface

The genesis of these notes was a series of four lectures given by the first author at the Tata Institute of Fundamental Research. It evolved into a joint project and contains many improvements and extensions on the material covered in the original lectures.

Let F be a finite extension of $\mathbb{Q}$, and E an elliptic curve defined over F. The fundamental idea of the Iwasawa theory of elliptic curves, which grew out of Iwasawa's basic work on the ideal class groups of cyclotomic fields, is to study deep arithmetic questions about E over F, via the study of coarser questions about the arithmetic of E over various infinite extensions of F. At present, we only know how to formulate this Iwasawa theory when the infinite extension is a p-adic Lie extension for a fixed prime number p. In fact, these notes will mainly discuss the simplest non-trivial example of the theory when our extension is the cyclotomic $\mathbb{Z}_p$-extension of F. However, we also make some comments about the theory when the extension is the field obtained by adjoining all p-power division points on E to F. When E does not admit complex multiplication, the Galois group of the extension of F obtained in this manner is an open subgroup of $GL_2(\mathbb{Z}_p)$, and it seems appropriate to refer to this situation as the GL_2 Iwasawa theory of E. We only discuss algebraic aspects of the Iwasawa theory of elliptic curves, both because of our desire to keep the notes short, and also because the analytic theory is in a much greater state of flux, in view of the recent deep contributions to it by K. Kato [K]. However, we have discussed in detail a number of numerical examples, which illustrate the general theory beautifully. Also, we suspect that the study of numerical examples will be vital in formulating some aspects of the general GL_2 Iwasawa theory of elliptic curves (for example, in studying the structure of the Selmer group as a module over the Iwasawa algebra). In addition, we have outlined some of the basic results in Galois cohomology which are used repeatedly in the study of the relevant Iwasawa modules. In the appendix, we briefly discuss some aspects of the cohomology of elliptic curves which seem to have been overlooked earlier and which turn out to be very useful in Iwasawa theory [C-S]. Finally, we refer the interested reader to

Greenberg's lecture notes [G2], which cover much the same ground as in these notes and more, but with a somewhat different view point.

In conclusion, the first author would like to thank the Tata Institute of Fundamental Research, Bombay, and the second author the Isaac Newton Institute of Mathematical Sciences, Cambridge, in particular the organisers of the Arithmetic Geometry Programme there in 1998, for generous support while these notes were being written.

J. Coates
R. Sujatha

Contents

Notation

We summarize here the notation used frequently in these notes. We write p for a prime number which will always be odd unless the contrary is stated. If L is a field, $\overline{L}$ will denote the separable closure of L. For each integer $m \geq 1$, μ_m will denote the group of m-th roots of unity in $\overline{L}$, viewed as a Galois module. If K is a Galois extension of L, we write $G(K/L)$ for the Galois group of K over L. Let μ_{p^∞} denote the group of all p-power roots of unity. We write F for a finite extension of $\mathbb{Q}$, which will nearly always be the base field over which we are working. If v is any place of F, F_v will denote the completion of F at v. We let K_∞ denote the cyclotomic $\mathbb{Z}_p$-extension of F, and put $\Gamma = G(K_\infty/F)$. Throughout, S will denote a finite set of non-archimedean primes of F, which will always be assumed to contain all primes of F above p. Let F_S denote the maximal extension of F unramified outside S and the archimedean primes of F. Put $G_S = G(F_S/F)$. Since S contains all primes of F dividing p, we have $K_\infty \subset F_S$, and we write $G_{S,\infty} = G(F_S/K_\infty)$. Throughout, E will denote an elliptic curve, which will usually be taken to be defined over the number field F. If L is an extension field of F, we write $E(L)$ for the group of L-rational points of F. If M is an abelian group, and m is an integer ≥ 1, we write M_m for the subgroup of elements of M which are annihilated by m. Put

$$M(p) = \bigcup_{n \geq 1} M_{p^n}, \quad T_p(M) = \varprojlim M_{p^n}.$$

Let M_{tors} denote the torsion subgroup of M. We write $M^* = \varprojlim M/p^n M$ for the p-adic completion of M. If M is p-primary (i.e., $M = M(p)$), we write M_{div} for the maximal p-divisible subgroup of M. When M is a discrete p-primary abelian group or a compact pro-p-abelian group, we let $\hat{M} = \mathrm{Hom}(M, \mathbb{Q}_p/\mathbb{Z}_p)$ (here Hom denotes the group of continuous homomorphisms) for the Pontryagin dual of M. If $\phi : M_1 \to M_2$ is a continuous homomorphism, we write $\hat{\phi} : \hat{M}_2 \to \hat{M}_1$ for the dual homomorphism. Let G be a profinite group, and M a G-module. As usual, M^G will be the subgroup of elements fixed by G, and M_G will be the largest quotient of M on which G acts trivially. If M is finite and p-primary, and μ_{p^∞} is also a G-module, we define $M^D = \mathrm{Hom}(M, \mu_{p^\infty})$. We endow M^D with the usual G-action, namely $(\sigma f)(x) = \sigma f(\sigma^{-1}x)$ for $\sigma \in G$ and $x \in M$. Similarly, if M is any p-primary G-module, we endow $\hat{M}$ with the G-action $(\sigma f)(x) = f(\sigma^{-1}x)$ for $\sigma \in G$ and $x \in M$. If M is a discrete G-module, we write $H^i(G, M)$ for the cohomology groups of G formed with continuous cochains. If L is a field, and $G = G(\overline{L}/L)$, we write as usual $H^i(L, M)$ rather than $H^i(G, M)$.

Chapter 1

Basic Results from Galois Cohomology

In this chapter, we gather all the results from Galois cohomology that will be used in the subsequent chapters. In particular, we discuss the classical long exact sequence due to Poitou-Tate, which lies at the heart of many calculations in Galois cohomology. We also derive a canonical modification of it, due originally to Cassels, who first introduced it in the special case of elliptic curves.

PRELIMINARY RESULTS

1.1. This paragraph recalls various pairings and duality results that will be used later.

Weil pairing: If L is any field of characteristic zero, and E/L is an elliptic curve, there is a non-degenerate, alternate, bilinear pairing [Ta]

$$w_m : E_m \times E_m \to \mu_m$$

where $E_m = E(\overline{L})_m$ and μ_m is the multiplicative group of m-th roots of unity. In fact, this is a particular case of such a pairing between A_m and A_m^t, where A/L is an arbitrary abelian variety and A^t its dual abelian variety.

Tate duality: Assume now that p is any prime, and L is a finite extension of $\mathbb{Q}_p$. Let E be an elliptic curve defined over L and let $E(L)$ be the group of L-rational points of E. The Galois cohomology group $H^1(L, E) = H^1(L, E(\overline{L}))$ classifies principal homogeneous spaces of E over L. There exists a non-degenerate bilinear pairing due to Tate (see [Si])

$$\Phi : E(L) \times H^1(L, E) \to \mathbb{Q}/\mathbb{Z}$$

such that Φ induces a duality of locally compact groups, when $E(L)$ is give the v-adic topology and $H^1(L, E)$ is endowed with the discrete topology.

3

Let L^{nr} denote the maximal unramified extension of L, and $E_0(L)$ be the subgroup of $E(L)$ consisting of elements with non-singular reduction [Si, p.173]. We shall need the well-known fact (cf. [Mc]) that under this dual pairing, the subgroup

$$W_v = H^1(G(L^{nr}/L), E(L^{nr})),$$

is the exact orthogonal complement of $E_0(L)$.

Tate local duality for finite Galois modules: Recall that $H^i(L, M) = 0$ except when $i = 0, 1, 2$, where M is any discrete $G(\overline{L}/L)$ module. If M is a finite p-primary group which is also a $G(\overline{L}/L)$-module, then for $0 \leq i \leq 2$, there is a canonical dual pairing [Se]

$$H^i(L, M) \times H^{2-i}(L, M^D) \to \mathbb{Q}_p/\mathbb{Z}_p$$

where we have $M^D = \mathrm{Hom}(M, \mu_{p^\infty})$, or equivalently, there is a canonical isomorphism

$$\widehat{H^i(L, M)} \simeq H^{2-i}(L, M^D).$$

We shall often tacitly treat this canonical isomorphism as an identification.

Local Euler-Poincaré characteristic: Put $G = G(\overline{L}/L)$. Given a finite G-module M, the groups $H^r(L, M)$ are finite for all r and trivial for $r \geq 2$, since the cohomological dimension of G is 2 [Se]. The *Euler characteristic* of M is defined as

$$\chi(G, M) = \#\, H^0(L, M).\#\, H^2(L, M)/\#\, H^1(L, M).$$

A basic result of Tate [Mi, p. 38] states that

$$\chi(G, M) = 1/[R : mR],$$

where R is the ring of integers in L, and $m = \#\, M$.

Global Euler-Poincaré characteristic: Let F be a finite extension of $\mathbb{Q}$ and v an archimedean place of F. Put

$$|\ \ |_v = \begin{cases} \text{ordinary absolute value if } v \text{ is real,} \\ \text{square of ordinary absolute value if } v \text{ is complex.} \end{cases}$$

If M is a finite p-primary G_S-module with $p \neq 2$, then the cohomology groups $H^i(G_S, M)$ are finite for all i and trivial for $i \geq 3$ [Ha, p. 21]. The *Euler characteristic* of M is defined as

$$\chi(G_S, M) = \#\, H^0(G_S, M).\#\, H^2(G_S, M)/\#\, H^1(G_S, M).$$

A well-known theorem of Tate [Ha, p. 30], [Mi, p. 82] gives the equality

$$\chi(G_S, M) = \prod_{v \mid \infty} (\# \, H^0(F_v, M) / \mid \#M \mid_v).$$

Kummer sequence: Let G be a profinite group, and let A be a discrete G-module which is divisible by p. The short exact sequence

$$0 \longrightarrow A_{p^n} \longrightarrow A \xrightarrow{p^n} A \longrightarrow 0$$

induces the following short exact sequence in Galois cohomology,

$$0 \longrightarrow A/p^n A \longrightarrow H^1(G, A_{p^n}) \longrightarrow H^1(G, A)_{p^n} \longrightarrow 0,$$

and this exact sequence is usually referred to as the Kummer sequence associated to A.

The results listed in the above paragraph will often be tacitly used in the subsequent sections. In Chapters 4 and 5, the duality results and the various pairings will be used explicitly in computations.

Poitou-Tate sequence

1.2. The next sequence we want to discuss is the celebrated Poitou-Tate sequence. As before, let F be a number field and M a finite p-primary G_S-module. We assume throughout this section that p is odd, to avoid a discussion of the role of the infinite primes of F. For each $v \in S$, there are restriction maps

$$\gamma_{v,i} : H^i(G_S, M) \to H^i(F_v, M)$$

for $i \geq 0$ which are induced by the composition of maps

$$G(\overline{F_v}/F_v) \longrightarrow\!\!\!\!\!\rightarrow G(F_{S,w}/F_v) \ensuremath{\lhook\joinrel\longrightarrow} G_S$$

where w denotes some prime of F_S above v, and $F_{S,w}$ is the union of the completions at w of the finite extensions of F contained in F_S. These restriction maps give canonical homomorphisms

$$\alpha_M : H^0(G_S, M) \to \bigoplus_{v \in S} H^0(F_v, M),$$

$$\beta_M : H^1(G_S, M) \to \bigoplus_{v \in S} H^1(F_v, M),$$

$$\gamma_M : H^2(G_S, M) \to \bigoplus_{v \in S} H^2(F_v, M),$$

where clearly α_M is injective. Recall that the groups $H^i(G_S, M)$ are finite, for all $i \geq 0$, with $H^i(G_S, M) = 0$ for all $i \geq 3$. Using Tate duality, we also have the canonical dual homomorphisms

$$\widehat{\alpha_M} : \bigoplus_{v \in S} H^2(F_v, M^D) \to H^0\widehat{(G_S, M)},$$

$$\widehat{\beta_M} : \bigoplus_{v \in S} H^1(F_v, M^D) \to H^1\widehat{(G_S, M)},$$

$$\widehat{\gamma_M} : \bigoplus_{v \in S} H^0(F_v, M^D) \to H^2\widehat{(G_S, M)},$$

with $\widehat{\alpha_M}$ surjective.

We thus have three sequences

$$0 \longrightarrow H^0(G_S, M) \xrightarrow{\alpha_M} \bigoplus_{v \in S} H^0(F_v, M) \xrightarrow{\widehat{\gamma_{M^D}}} H^2\widehat{(G_S, M^D)},$$

$$H^1(G_S, M) \xrightarrow{\beta_M} \bigoplus_{v \in S} H^1(F_v, M) \xrightarrow{\widehat{\beta_{M^D}}} H^1\widehat{(G_S, M^D)},$$

$$H^2(G_S, M) \xrightarrow{\gamma_M} \bigoplus_{v \in S} H^2(F_v, M) \xrightarrow{\widehat{\alpha_{M^D}}} H^0\widehat{(G_S, M^D)} \longrightarrow 0.$$

The homomorphisms

$$H^r(G_S, M) \to \bigoplus_{v \in S} H^r(F_v, M)$$

have the property that Ker β_M and Ker $\widehat{\alpha_M}$ are finite. Further, there is a canonical non-degenerate pairing

$$\text{Ker } \beta_M \times \text{Ker } \widehat{\alpha_M} \to \mathbb{Q}_p/\mathbb{Z}_p.$$

Even the definition of this pairing is rather involved. That it is non-degenerate follows using deep results of class-field theory (cf. [Ha, §1.4]).

Granting these results, they can be used to construct a connecting homomorphism

$$\delta_M : H^2\widehat{(G_S, M^D)} \to H^1(G_S, M).$$

The Poitou-Tate theorem connects the three sequences above using the connecting homomorphism δ_M. We refer to [Ha], [P] for more details and simply state the theorem.

1.3. Theorem. (Poitou-Tate) *There is a long exact sequence*

$$0 \longrightarrow H^0(G_S, M) \xrightarrow{\alpha_M} \underset{v \in S}{\oplus} H^0(F_v, M) \xrightarrow{\widehat{\gamma_{M^D}}} H^2\widehat{(G_S, M^D)}$$

$$\downarrow \delta_M$$

$$H^1\widehat{(G_S, M^D)} \xleftarrow{\widehat{\beta_{M^D}}} \underset{v \in S}{\oplus} H^1(F_v, M) \xleftarrow{\beta_M} H^1(G_S, M)$$

$$\downarrow \widehat{\delta_{M^D}}$$

$$H^2(G_S, M) \xrightarrow{\gamma_M} \underset{v \in S}{\oplus} H^2(F_v, M) \longrightarrow H^0\widehat{(G_S, M^D)} \longrightarrow 0.$$

CASSELS-POITOU-TATE SEQUENCE

1.4. We next discuss a modification of the Poitou-Tate sequence, originally due to Cassels [Ca]. Cassels considered the case of $M = E_{p^n}$, the p^n-torsion points of $E(\overline{F})$, where E is an elliptic curve defined over F. We treat a more general situation than that considered by Cassels. From the Poitou-Tate sequence, we have Im $\delta_M = $ Ker β_M. We suppose that for all $v \in S$, we are given a subgroup $W_v(M)$ of $H^1(F_v, M)$. We define $W_v(M^D)$ to be the orthogonal complement of $W_v(M)$ in the Tate pairing

$$H^1(F_v, M) \times H^1(F_v, M^D) \to \mathbb{Q}_p/\mathbb{Z}_p.$$

We can thus canonically identify the dual of $H^1(F_v, M)/W_v(M)$ with $W_v(M^D)$ and vice-versa, which we do in what follows. Let ϕ_M be the composite

$$\phi_M : H^1(G_S, M) \xrightarrow{\beta_M} \underset{v \in S}{\oplus} H^1(F_v, M) \longrightarrow \underset{v \in S}{\oplus} H^1(F_v, M)/W_v(M)$$

where the second map is the natural projection. We also define maps

$$\rho_M : \text{Ker } \phi_M \to \underset{v \in S}{\oplus} W_v(M),$$

with ρ_M being the restriction of β_M, and

$$\eta_M : H^2\widehat{(G_S, M^D)} \to \text{Ker } \phi_M.$$

The map η_M exists because Im $\delta_M = $ Ker $\beta_M \subseteq $ Ker ϕ_M. Observe that the sequence

$$H^2\widehat{(G_S, M^D)} \xrightarrow{\eta_M} \text{Ker } \phi_M \xrightarrow{\rho_M} \underset{v \in S}{\oplus} W_v(M)$$

is exact.

1.5. Theorem. (Cassels-Poitou-Tate) *There is a canonical exact sequence*

$$0 \longrightarrow \mathrm{Ker}\ \phi_M \longrightarrow H^1(G_S, M) \xrightarrow{\ \phi_M\ } \bigoplus_{v \in S} H^1(F_v, M)/W_v(M)$$

$$\Big\downarrow \widehat{\rho_{M^D}}$$

$$\bigoplus_{v \in S} H^2(F_v, M) \xleftarrow{\ \gamma_M\ } H^2(G_S, M) \xleftarrow{\ \widehat{\eta_{M^D}}\ } \mathrm{Ker}\ \widehat{\phi_{M^D}}$$

$$\Big\downarrow \widehat{\alpha_{M^D}}$$

$$H^0\widehat{(G_S, M^D)} \longrightarrow 0.$$

Proof. We only have to show that the sequence is exact at the term $\bigoplus_{v \in S} H^1(F_v, M)/W_v(M)$, or equivalently that the dual sequence

$$\mathrm{Ker}\ \phi_{M^D} \xrightarrow{\ \rho_{M^D}\ } \bigoplus_{v \in S} W_v(M^D) \xrightarrow{\ \widehat{\phi_M}\ } H^1\widehat{(G_S, M)}$$

is exact at the term in the middle. The commutative diagram

$$H^1(G_S, M^D)$$

$$\Big\downarrow \beta_{M^D} \qquad \searrow^{\ \phi_{M^D}}$$

$$\bigoplus_{v \in S} H^1(F_v, M^D) \longrightarrow \bigoplus_{v \in S} H^1(F_v, M^D)/W_v(M^D)$$

yields the exact sequence

$$0 \longrightarrow \mathrm{Ker}\ \beta_{M^D} \longrightarrow \mathrm{Ker}\ \phi_{M^D} \xrightarrow{\ \rho_{M^D}\ } \bigoplus_{v \in S} W_v(M^D)$$

$$\xrightarrow{\ \lambda_{M^D}\ } \mathrm{Coker}(\beta_{M^D}) \longrightarrow \mathrm{Coker}(\phi_{M^D}) \longrightarrow 0.$$

Hence $\mathrm{Im}\ \rho_{M^D} = \mathrm{Ker}\ \lambda_{M^D}$. On the other hand, the original Poitou-Tate sequence yields a canonical inclusion

$$\widehat{\beta_M} : \mathrm{Coker}(\beta_{M^D}) \hookrightarrow H^1\widehat{(G_S, M)}.$$

Moreover, the map $\widehat{\phi_M}$ is the composite

$$W_v(M^D) \xrightarrow{\ \lambda_{M^D}\ } \mathrm{Coker}(\beta_{M^D}) \xrightarrow{\ \widehat{\beta_M}\ } H^1\widehat{(G_S, M)}$$

and therefore $\mathrm{Ker}\ \lambda_{M^D} = \mathrm{Ker}\ \widehat{\phi_M}$. Hence $\mathrm{Im}\ \rho_{M^D} = \mathrm{Ker}\ \widehat{\phi_M}$, as required.

1.6. We comment briefly on the various possible choices of the subgroups $W_v(M)$.

(i): $v \neq p$. There are two natural choices which occur in literature. One is the case $W_v(M) = 0$ and this is the case that is exploited by Wiles [Wi]. Note that with this choice, $W_v(M^D) = H^1(F_v, M^D)$. Another case is the subgroup $W_v(M) = H^1(G_v/I_v, M^{I_v})$, where I_v is the inertial subgroup of $G_v = G(\overline{F_v}/F_v)$. The subgroup $W_v(M)$ is the group of unramified cocycles. A basic fact that we mention in this context is that in this case, $W_v(M^D) = H^1(G_v/I_v, (M^D)^{I_v})$. We refer to [Se, II 5.5] for a proof, remarking that the assumption there of M being unramified is unnecessary.

(ii): In this case, we choose $M = E_{p^n}$, the p^n-torsion points over $\overline{F}$ of an elliptic curve E defined over F, (more generally, one can consider an arbitrary abelian variety). Then the Kummer sequence gives a homomorphism

$$E(F_v)/p^n E(F_v) \xhookrightarrow{\kappa_{E,p^n}} H^1(F_v, E_{p^n})$$

and one considers the subgroup $W_v(E_{p^n}) = \mathrm{Im}(\kappa_{E,p^n})$.

If one considers an arbitrary abelian variety A, then there is the non-degenerate Weil pairing

$$A_{p^n} \times A^t_{p^n} \to \mu_{p^n}$$

where A^t is the dual abelain variety. Hence if $M = A_{p^n}$, then $M^D = A^t_{p^n}$. A basic fact (due to Tate) is that in this situation, $W_v(M^D) = W_v(A^t_{p^n}) = \mathrm{Im}(\kappa_{A^t,p^n})$ where κ_{A^t,p^n} is the corresponding map arising from the Kummer sequence for A^t. A generalization of these for arbitrary M has been exploited by Bloch-Kato [B-K], in their conjectures about the Tamagawa numbers of motives.

CASSELS-POITOU-TATE SEQUENCE FOR ELLIPTIC CURVES

1.7. We now discuss how the Cassels-Poitou-Tate sequence can be refined further when one considers elliptic curves over F. In this case it is well-known that E_{p^n} is a G_S-module [Si], where $G_S = G(F_S/F)$, and S is any finite set of primes containing all primes dividing p and all primes where E has bad reduction. We let $M = E_{p^n}$ and note that $M^D = E_{p^n}$ since E is self-dual as an abelian variety. For the subgroup $W_v(M) \subseteq H^1(F_v, M) = H^1(F_v, E_{p^n})$, we choose $W_v(E_{p^n}) = \mathrm{Im}(\kappa_{E,p^n})$. By the remarks in the previous paragraph, we have $W_v(E_{p^n}^D) = \mathrm{Im}(\kappa_{E,p^n})$. Further, the group $H^1(G_S, E_{p^n}) \subseteq H^1(F, E_{p^n})$ and the kernel of the natural map

$$\lambda_S : H^1(G_S, E_{p^n}) \to \bigoplus_{v \in S} H^1(F_v, E)_{p^n} \simeq \bigoplus_{v \in S} H^1(F_v, E_{p^n}) / \mathrm{Im}(\kappa_{E,p^n})$$

is in fact the classical Selmer group of E relative to p^n, which we denote by $\mathcal{S}(E/F, p^n)$ or simply $\mathcal{S}(E, p^n)$. The latter isomorphism above follows from the Kummer sequence applied to E over F_v for $v \in S$. We recall [Si, Chapter X, §4] that $\mathcal{S}(E/F, p^n)$ is defined by the exactness of

$$0 \longrightarrow \mathcal{S}(E/F, p^n) \longrightarrow H^1(F, E_{p^n}) \longrightarrow \prod_v H^1(F_v, E)$$

where v now ranges over all non-archimedean places of F, and that we have the exact sequence

$$0 \longrightarrow E(F)/p^n \longrightarrow \mathcal{S}(E/F, p^n) \longrightarrow \text{III}(E/F)_{p^n} \longrightarrow 0 \tag{1}$$

where $\text{III}(E/F)_{p^n}$ or simply $\text{III}(E)_{p^n}$ denotes the p^n-torsion subgroup of the Tate-Shafarevich group $\text{III}(E)$ of E. In particular, we see that with this choice of the subgroup $W_v(E_{p^n})$, the group Ker λ_S is in fact independent of the choice of S. The Cassels-Poitou-Tate sequence takes the form

$$0 \longrightarrow \mathcal{S}(E, p^n) \longrightarrow H^1(G_S, E_{p^n}) \longrightarrow \bigoplus_{v \in S} H^1(F_v, E)_{p^n}$$

$$\longrightarrow \widehat{\mathcal{S}(E, p^n)} \longrightarrow H^2(G_S, E_{p^n}) \longrightarrow \bigoplus_{v \in S} H^2(F_v, E_{p^n}) \tag{2}$$

$$\longrightarrow \widehat{E_{p^n}(F)} \longrightarrow 0.$$

This is the sequence that was discovered by Cassels [Ca]. We can now vary n and pass to the limit as $n \to \infty$. This can be done in two different ways. First, we can pass to the inductive limit using the inclusion

$E_{p^n} \hookrightarrow E_{p^{n+1}}$. Letting $E_{p^\infty} = \varinjlim E_{p^n}$, the above equation then becomes

$$0 \longrightarrow \mathcal{S}(E/F) \longrightarrow H^1(G_S, E_{p^\infty}) \longrightarrow \bigoplus_{v \in S} H^1(F_v, E)(p)$$

$$\longrightarrow \widehat{\mathfrak{S}(E/F)} \longrightarrow H^2(G_S, E_{p^\infty}) \longrightarrow \bigoplus_{v \in S} H^2(F_v, E_{p^\infty}) \qquad (3)$$

$$\longrightarrow \widehat{T_p E(F)} \longrightarrow 0.$$

Here

$$\mathcal{S}(E/F) = \varinjlim \mathcal{S}(E, p^n) \subseteq \varinjlim H^1(G_S, E_{p^n}),$$

while

$$\varinjlim H^1(G_S, E_{p^n}) = H^1(G_S, E_{p^\infty}) \subseteq H^1(F, E_{p^\infty})$$

and

$$\mathfrak{S}(E/F) = \varprojlim_{n} \mathcal{S}(E, p^n) \subseteq H^1(F, T_p E).$$

Further, the Selmer group $\mathcal{S}(E/F)$ is a discrete subgroup of $H^1(F, E_{p^\infty})$ while $\mathfrak{S}(E/F)$ is a compact subgroup of $H^1(F, T_p E)$. By Tate duality, the group $H^2(F_v, E_{p^\infty}) = \varinjlim_{n} H^2(F_v, E_{p^n})$ is dual to $\varprojlim_{n} H^0(F_v, E_{p^n})$, where the inverse limit is with respect to "multiplication by p" map from $E_{p^{n+1}} \to E_{p^n}$. But $E_{p^n}(F_v)$ is a finite group and hence the inverse limit above is trivial. Therefore we in fact have an exact sequence

$$0 \longrightarrow \mathcal{S}(E/F) \longrightarrow H^1(G_S, E_{p^\infty}) \longrightarrow \bigoplus_{v \in S} H^1(F_v, E)(p)$$

$$\qquad (4)$$

$$\longrightarrow \widehat{\mathfrak{S}(E/F)} \longrightarrow H^2(G_S, E_{p^\infty}) \longrightarrow 0.$$

The discrete Selmer group $\mathcal{S}(E/F)$ and the compact Selmer group $\mathfrak{S}(E/F)$ are related as follows.

1.8. Lemma. *There is an exact sequence*

$$0 \longrightarrow E_{p^\infty}(F) \longrightarrow \mathfrak{S}(E/F) \longrightarrow T_p \mathcal{S}(E/F) \longrightarrow 0.$$

Proof. Choose $n >> 0$ so that p^n kills $E_{p^\infty}(F)$. Consider the exact sequence of $G(\overline{F}/F)$-modules

$$0 \longrightarrow E_{p^n}(\overline{F}) \longrightarrow E_{p^\infty}(\overline{F}) \xrightarrow{\ p^n\ } E_{p^\infty}(\overline{F}) \longrightarrow 0.$$

This induces the long exact sequence

$$0 \longrightarrow E_{p^\infty}(F) \longrightarrow H^1(F, E_{p^\infty}) \longrightarrow H^1(F, E_{p^\infty})_{p^n} \longrightarrow 0$$

and also an exact sequence

$$0 \longrightarrow E_{p^\infty}(F) \longrightarrow \mathcal{S}(E/F, p^n) \longrightarrow \mathcal{S}(E/F)_{p^n} \longrightarrow 0.$$

Taking inverse limits and observing that the groups $\mathcal{S}(E/F)_{p^n}$ are finite [Si], we get an exact sequence

$$0 \longrightarrow E_{p^\infty}(F) \longrightarrow \mathfrak{S}(E/F) \longrightarrow T_p\mathcal{S}(E/F) \longrightarrow 0$$

and the lemma is proved.

$$\square$$

We now record some important consequences of the Cassels-Poitou-Tate sequence.

1.9. Proposition. *Assume that both $E(F)$ and $\text{III}(E/F)(p)$ are finite. Then*

(i) $H^2(G(F_S/F), E_{p^\infty}) = 0$.

(ii) *If λ_F is the natural map*

$$\lambda_F : H^1(G(F_S/F), E_{p^\infty}) \to \bigoplus_{v \in S} H^1(F_v, E)(p),$$

then Coker λ_F *is finite and* $\#(\text{Coker } \lambda_F) = \#(E(F)(p))$.

Proof. The exact sequence

$$0 \longrightarrow E(F)/p^n \longrightarrow \mathcal{S}(E, p^n) \longrightarrow \text{III}(E)_{p^n} \longrightarrow 0$$

gives an exact sequence (on taking $\varprojlim_{n}$ and noting that $\text{III}(E)_{p^n}$ is finite)

$$0 \longrightarrow E(F)^* \longrightarrow \mathfrak{S}(E/F) \longrightarrow \dot{T}_p\text{III}(E/F) \longrightarrow 0.$$

Since $\text{III}(E)(p)$ is finite by assumption, we have $T_p(\text{III}(E)) = 0$ and hence $E(F)^* = \mathfrak{S}(E/F)$. The exact sequence (4) therefore gives

$$H^1(G(F_S/F), E_{p^\infty}) \xrightarrow{\ \lambda_F\ } \underset{v \in S}{\oplus}\, H^1(F_v, E)(p) \xrightarrow{\ \theta_F\ } \widehat{E(F)^*}$$

$$\xrightarrow{\hspace{3cm}} H^2(G(F_S/F), E_{p^\infty}) \xrightarrow{\hspace{2cm}} 0.$$

But $E(F)$ finite implies that $E(F)^* = E(F)(p)$ and $\widehat{\theta_F}$ is the map (use Tate duality)

$$\widehat{\theta_F} : E(F)(p) \to \underset{v \in S}{\oplus}\, E(F_v)^*$$

given by the natural inclusion. Since $\widehat{\theta_F}$ is injective, it is plain that θ_F is surjective and the proposition follows.

$\square$

1.10. Remark. As a special case, consider a modular elliptic curve $E/\mathbb{Q}$ such that the L-value $L(E,1) \neq 0$. Then, by the deep theorem of Kolyvagin [K], the hypotheses of Proposition 1.9 are satisfied for all primes p. Hence we conclude that for all $p \neq 2$, $H^2(G(\mathbb{Q}_S/\mathbb{Q}), E_{p^\infty}) = 0$ and Coker $\lambda_\mathbb{Q}$ is finite of order $\#\, E(\mathbb{Q})(p)$. We have

$$H^2(G(\mathbb{Q}_S/\mathbb{Q}), E_{p^\infty}) = 0$$

and Coker $\phi_\mathbb{Q}$ is finite of order $\#\, E(\mathbb{Q})(p)$.

We next establish a formula which is reminiscent of Tamagawa measures of $E(F_v)$ for v a finite place of F.

1.11. Lemma. *Let E/F be an elliptic curve and v be a prime of F that does not lie above p. Then $\#\, H^1(F_v, E)(p) \sim c_v/L_v(E,1)$ where $u \sim v$ signifies that u/v is a p-adic unit.*

Proof. Let $E_0(F_v) \subseteq E(F_v)$ be the subgroup which consists of points whose reduction modulo v lies in the subgroup $\tilde{E}_{ns}(k_v)$. Here k_v is the residue field and $\tilde{E}$ is the reduced curve while $\tilde{E}_{ns}$ denotes the subset consisting of the non-singular points of the reduced curve. Let $B_v = E(F_v)/E_0(F_v)$ and $c_v = \#B_v$. Since B_v is finite [Si, Chapter VII, §6], on taking p-adic completions of the exact sequence

$$0 \longrightarrow E_0(F_v) \longrightarrow E(F_v) \longrightarrow B_v \longrightarrow 0,$$

we obtain the exact sequence

$$0 \longrightarrow E_0(F_v)^* \longrightarrow E(F_v)^* \longrightarrow B_v^* \longrightarrow 0.$$

Consider the exact sequence

$$0 \longrightarrow E_1(F_v) \longrightarrow E_0(F_v) \longrightarrow \tilde{E}_{ns}(k_v) \longrightarrow 0.$$

The subgroup $E_1(F_v)$ can be identified with the points on the formal group of E at v. Now v does not divide p implies that p is an automorphism of $E_1(F_v)$ and hence $E_1(F_v)^* = 0$. Therefore $E_0(F_v)^* \simeq \tilde{E}_{ns}(k_v)^*$. But it is well-known [Si] that $\# \tilde{E}_{ns}(k_v) = N(v)/L_v(E,1)$ and therefore $\# E(F_v)^* \sim c_v/L_v(E,1)$. The lemma now follows from Tate duality **(1.1)**.

□

We will need the following useful lemma, part of which has already been referred to in **(1.7)**.

1.12. Lemma. *Let L be a non-archimedean local field and A/L an abelian variety. Then $H^2(L, A) = 0$.*

Proof. Since the cohomological dimension of a local field is 2, the Kummer sequence **(1.1)** gives a surjection

$$H^2(L, A_{p^\infty}) \to H^2(L, A)(p)$$

for any prime p. We prove that $H^2(L, A_{p^\infty}) = 0$. By Tate duality **(1.1)**, the groups $H^2(L, A_{p^n})$ are dual to $A(L)_{p^n}$ for all n. Thus, on taking limits, we see that $H^2(L, A_{p^\infty})$ is dual to the Tate module $T_p(A(L))$. But this latter group is zero as the p-primary subgroup of $A(L)$ is finite. Since $H^2(L, A)$ is a torsion group with $H^2(L, A)(p) = 0$ for any prime p, the lemma is proved.

□

The following is left as an exercise to the reader.

1.13. Exercise. Suppose E/F is an elliptic curve such that $E(F)$ and $\text{III}(E/F)$ are both finite. Let T be a finite non-empty set of primes of F not containing any prime lying above p and define

$$\text{III}_T(E/F) = \text{Ker}(H^1(F, E) \to \prod_{v \notin T} H^1(F_v, E)).$$

Show that the index

$$[\text{Ш}_T(E/F)(p) : \text{Ш}(E/F)(p)] \sim \#\,(E(F)(p))^{-1} \cdot \prod_{v \in T} c_v . L_v(E,1)^{-1}.$$

(*Hint:* Use Lemma 1.11 and the Cassels-Poitou-Tate sequence).

We do not know the answer to the following question.

1.14. Question. Is the above formula for the index valid if we do not assume that $E(F)$ and $\text{Ш}(E/F)$ are finite?

Chapter 2

The Iwasawa Theory of the Selmer Group

INTRODUCTION

2.1. Throughout this chapter, F will denote a finite extension of $\mathbb{Q}$, and E will denote an elliptic curve defined over F. The main goal of the arithmetic of elliptic curves is to provide a p-adic approach to the conjecture of Birch and Swinnerton-Dyer for E over the base field F. We begin by describing a rather general setting in which the ideas of Iwasawa theory can be fruitfully applied. Let H_∞ denote an infinite Galois extension of F whose Galois group $\Omega = G(H_\infty/F)$ is a p-adic Lie group of positive dimension. In these notes, we shall mainly be concerned with the case in which H_∞ is the cyclotomic $\mathbb{Z}_p$-extension of F. We recall that the cyclotomic $\mathbb{Z}_p$-extension of F is the fixed field of the torsion subgroup of the Galois group of $F(\mu_{p^\infty})$ over F, where μ_{p^∞} denotes the group of all p-power roots of unity. It is the unique subextension of $F(\mu_{p^\infty})$ over F whose Galois group is topologically isomorphic to the additive group of the ring $\mathbb{Z}_p$ of p-adic integers. However, we shall also discuss from time to time the case in which $H_\infty = F(E_{p^\infty})$, where E_{p^∞} denotes the p-power torsion points of $E(\overline{F})$. Many other choices of the field H_∞ are also possible. It is no exaggeration to say that one reason why the Iwasawa theory of elliptic curves is so rich in problems is because we are free to make many choices of the field H_∞.

By direct analogy with the situation over F discussed in Chapter 1, we define the Selmer group $\mathcal{S}(E/H_\infty)$ of E over H_∞ by

$$\mathcal{S}(E/H_\infty) = \mathrm{Ker}(H^1(H_\infty, E_{p^\infty}) \to \prod_w H^1(H_{\infty,w}, E)).$$

Here w runs over all finite primes of H_∞, and, as usual for infinite extensions, $H_{\infty,w}$ denotes the union of the completions at w of all finite extensions of F contained in H_∞. It is also easy to see that

$$\mathcal{S}(E/H_\infty) = \varinjlim \mathcal{S}(E/L),$$

where L runs over all finite extensions of F contained in H_∞, $\mathcal{S}(E/L)$ denotes the Selmer group of E over L as defined in Chapter 1, and the inductive limit is taken with respect to the restriction maps. Again, it follows immediately from Kummer theory for E over H_∞ that we have the exact sequence

$$0 \longrightarrow E(H_\infty) \otimes \mathbb{Q}_p/\mathbb{Z}_p \longrightarrow \mathcal{S}(E/H_\infty) \longrightarrow \text{III}(E/H_\infty)(p) \longrightarrow 0,$$

where the Tate-Shafarevich group $\text{III}(E/H_\infty)$ of E over H_∞ is defined by

$$\text{III}(E/H_\infty) = \text{Ker}(H^1(H_\infty, E) \to \prod_w H^1(H_{\infty,w}, E)),$$

and w again runs over all finite primes of H_∞.

The central idea of Iwasawa theory is to oberve that the Galois group Ω of H_∞ over F has a natural left action on $\mathcal{S}(E/H_\infty)$, and to use this Ω-module structure to study $\mathcal{S}(E/F)$, and, in particular, to relate $\mathcal{S}(E/F)$ to L-functions in the spirit of the conjecture of Birch and Swinnerton-Dyer. The action of Ω on $\mathcal{S}(E/H_\infty)$ is the obvious one arising from the action of Ω on $H^1(H_\infty, E_{p^\infty})$. This action makes $\mathcal{S}(E/H_\infty)$ into a discrete p-primary Ω-module. It will often be more convenient to study its compact dual $\widehat{\mathcal{S}(E/H_\infty)} = \text{Hom}(\mathcal{S}(E/H_\infty), \mathbb{Q}_p/\mathbb{Z}_p)$, which is endowed with the left action of Ω given by $(\sigma f)(x) = f(\sigma^{-1}x)$ for f in $\widehat{\mathcal{S}(E/H_\infty)}$ and σ in Ω.

Clearly $\mathcal{S}(E/H_\infty)$ and $\widehat{\mathcal{S}(E/H_\infty)}$ are continuous modules over the ordinary group ring $\mathbb{Z}_p[\Omega]$ of Ω with coefficients in $\mathbb{Z}_p$. But, as Iwasawa was the first to exploit in the case of the cyclotomic theory, it is much more useful to view them as modules over a larger algebra, which we denote by $\Lambda(\Omega)$ and call the Iwasawa algebra of Ω, and which is defined by

$$\Lambda(\Omega) = \varprojlim \mathbb{Z}_p[\Omega/W],$$

where W runs over all open normal subgroups of Ω. The algebra $\Lambda(\Omega)$ has an interpretaion as the ring of measures on Ω with values in $\mathbb{Z}_p$, a fact which is very convenient for studying analysis on Ω and p-adic L-functions. A second, more algebraic, interpretation of $\Lambda(\Omega)$ is as follows.

If A is any discrete p-primary Ω-module, and $X = \operatorname{Hom}(A, \mathbb{Q}_p/\mathbb{Z}_p)$ is its Pontrjagin dual, then we have

$$A = \bigcup_W A^W, \qquad X = \varprojlim X_W,$$

where W again runs over all open normal subgroups of Ω, and X_W denotes the largest quotient of X on which W acts trivially. It is then clear how to extend by continuity the natural action of $\mathbb{Z}_p[\Omega]$ on A and X to an action of the whole Iwasawa algebra $\Lambda(\Omega)$.

THE FUNDAMENTAL DIAGRAM

2.2. The crucial ingredient in studying $\mathcal{S}(E/H_\infty)$ as a module over the Iwasawa algebra $\Lambda(\Omega)$ is the single natural commutative diagram given below. Because of its importance, we shall henceforth call it the *fundamental diagram*.

As in Chapter 1, let S be any finite set of primes of F which contains all the primes dividing p, and all primes where E has bad reduction. Let F_S denote the maximal extension of F which is unramified outside of S and all the archimedean primes of F, and let G_S be the corresponding Galois group. We assume that S also has the property that $H_\infty \subset F_S$ (this is automatically true for the two choices of H_∞ made in these notes). We set $G_{S,\infty} = G(F_S/H_\infty)$. We thus have the following tower of extensions:

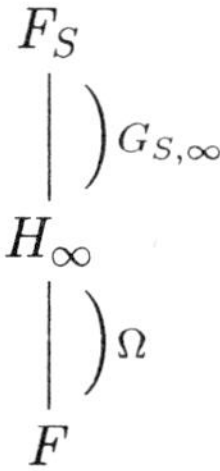

For any finite extension L of F contained in H_∞, we write

$$J_v(L) = \bigoplus_{w|v} H^1(L_w, E)(p) \tag{5}$$

where the direct sum is taken over all primes w of L lying above the prime v of F. We then define

$$J_v(H_\infty) = \varinjlim J_v(L) \tag{6}$$

where the inductive limit is taken with respect to the restriction maps, and L runs over all finite extensions of F contained in H_∞. Recall that E_{p^∞} denotes the G_S-module of all torsion points in $E(\overline{F})$ whose order is a power of p. Then, as is explained in Chapter 1, we have the exact sequence

$$0 \longrightarrow \mathcal{S}(E/F) \longrightarrow H^1(G_S, E_{p^\infty}) \overset{\lambda}{\longrightarrow} \underset{v \in S}{\oplus}\, J_v(F), \qquad (7)$$

where λ is the obvious localization map. There is an analogous exact sequence

$$0 \longrightarrow \mathcal{S}(E/H_\infty) \longrightarrow H^1(G_{S,\infty}, E_{p^\infty}) \overset{\lambda_\infty}{\longrightarrow} \underset{v \in S}{\oplus}\, J_v(H_\infty), \qquad (8)$$

which is obtained by taking the direct limit of the exact sequences (7) over all finite extensions of F contained in H_∞. We include the case $p = 2$, noting that in this case the corresponding Selmer groups are bigger than the classical ones because we have not imposed local conditions at the archimedean primes. Of course, the sequence (7) is a sequence of Ω-modules and on taking Ω-invariants we obtain the fundamental diagram

$$
\begin{array}{ccccccc}
0 & \longrightarrow & \mathcal{S}(E/H_\infty)^\Omega & \longrightarrow & H^1(G_{S,\infty}, E_{p^\infty})^\Omega & \overset{\phi_\infty}{\longrightarrow} & \underset{v \in S}{\oplus}\, J_v(H_\infty)^\Omega \\
 & & \big\uparrow{\scriptstyle \alpha} & & \big\uparrow{\scriptstyle \beta} & & \big\uparrow{\scriptstyle \gamma} \\
0 & \longrightarrow & \mathcal{S}(E/F) & \longrightarrow & H^1(G_S, E_{p^\infty}) & \overset{\lambda}{\longrightarrow} & \underset{v \in S}{\oplus}\, J_v(F),
\end{array}
\qquad (9)
$$

where the rows are exact, and the vertical arrows are the obvious restriction maps with $\gamma = \underset{v \in S}{\oplus}\, \gamma_v$. We emphasize that all of our subsequent arguments revolve around analyzing this diagram. In particular, the analysis of Ker γ and Coker γ is a purely local question whose answer at the primes v dividing p will depend on the nature of the reduction of E at v.

CYCLOTOMIC THEORY

2.3. We now turn to the case with which we will be mainly concerned in these notes, namely when H_∞ is the cyclotomic $\mathbb{Z}_p$-extension of the base field F. We write K_∞ instead of H_∞ in this case, and we also write Γ instead of Ω for the Galois group of K_∞ over F. Thus Γ is topologically isomorphic to $\mathbb{Z}_p$, and it is well-known (cf. [Se1],[W], see Appendix A.1) that the Iwasawa algebra $\Lambda(\Gamma)$ is isomorphic to the ring $\mathbb{Z}_p[[T]]$ of formal power series in an indeterminate T with coefficients in $\mathbb{Z}_p$. In fact, there is a unique isomorphism of topological $\mathbb{Z}_p$-algebras which maps any given topological generator of Γ to $1 + T$. Also, the structure theory of finitely generated modules over $\Lambda(\Gamma)$ is well-known, but we shall avoid using this as much as possible and for much of this chapter we will only need the fact that $\Lambda(\Gamma)$ has no zero divisors.

2.4. Lemma. *For every prime p, the dual Selmer group $\widehat{\mathcal{S}(E/K_\infty)}$ is finitely generated over $\Lambda(\Gamma)$.*

Proof. By a well-known variant of Nakayama's lemma ([Se1],[W]), it suffices to show that $\widehat{\mathcal{S}(E/K_\infty)}_\Gamma$ is finitely generated over $\mathbb{Z}_p$, or equivalently by duality, that $\mathcal{S}(E/K_\infty)^\Gamma$ has finite $\mathbb{Z}_p$-corank. But this latter statement is an easy consequence of the fundamental diagram (9) for the cyclotomic $\mathbb{Z}_p$-extension. Indeed, the fact that $\mathcal{S}(E/F)$ has finite $\mathbb{Z}_p$-corank follows easily from the well-known result that $H^1(G_S, E_{p^\infty})$ has finite $\mathbb{Z}_p$-rank (this can be easily deduced for example, from Tate's global Euler characteristic formula, cf. **(1.1)**). Also $\mathrm{Ker}\,\gamma$ has finite $\mathbb{Z}_p$-corank because it is dual, via Tate duality **(1.1)**, to a quotient of $\underset{v \in S}{\oplus}\, E(F_v)^*$. That $\underset{v \in S}{\oplus}\, E(F_v)^*$ is a finitely generated $\mathbb{Z}_p$-module follows from the well-known structure of $E(F_v)$ [Si, Chapter VII] involving the formal group. Further, in the fundamental diagram, the map β is surjective because Γ has cohomological dimension 1, and from the Hochschild-Serre spectral sequence, $\mathrm{Coker}\,\beta$ injects into $H^2(\Gamma, E_{p^\infty}(K_\infty)) = 0$. Thus the map α has cokernel of finite $\mathbb{Z}_p$-rank by the snake lemma, and so it follows that $\mathcal{S}(E/K_\infty)^\Gamma$ has finite $\mathbb{Z}_p$-corank, as required.

$\square$

Let $Q(\Gamma)$ denote the quotient field of $\Lambda(\Gamma)$. If X is a finitely generated $\Lambda(\Gamma)$-module, we recall that the $\Lambda(\Gamma)$-rank of X is defined to be the $Q(\Gamma)$-dimension of $X \otimes_{\Lambda(\Gamma)} Q(\Gamma)$. It is natural to ask what the $\Lambda(\Gamma)$-rank of the finitely generated $\Lambda(\Gamma)$-module $\widehat{\mathcal{S}(E/K_\infty)}$ is. The conjectural answer for this depends on the nature of the reduction of E at the places v of F deviding p. We recall that E is said to have *potential supersingular reduction* at a prime v if there exists a finite extension L of F_v such that E/L has good supersingular reduction. For example, the elliptic curve $y^2 = x^3 - x$ over $\mathbb{Q}$ has bad but potentially supersingular reduction, at the prime 2. Let $P_p(E/F)$ be the set of places v of F dividing p such that E has potential supersingular reduction at v. We say that E/F is of *supersingular type* at p if $P_p(E/F)$ is non-empty. We define

$$
r_p(E/F) = \begin{cases} \sum_{v \in P_p(E/F)} [F_v : Q_p] & \text{if } E/F \text{ is of supersingular type at } p, \\ 0 & \text{if } E/F \text{ is not of super-singular type at } p. \end{cases} \tag{10}
$$

The following conjecture is folklore:

2.5. Conjecture. *For every prime p, the $\Lambda(\Gamma)$-rank of $\mathcal{S}(\widehat{E/K_\infty})$ is equal to $r_p(E/F)$.*

The following partial result in the direction of this conjecture is well-known, and we shall sketch its proof later in this chapter.

2.6. Theorem. *For all primes p, the $\Lambda(\Gamma)$-rank of $\mathcal{S}(\widehat{E/K_\infty})$ is greater than or equal to $r_p(E/F)$.*

2.7. Examples. We shall illustrate our conjectures and results for the three elliptic curves of conductor 11 defined over $\mathbb{Q}$, namely

$$
\begin{aligned}
A_0 : y^2 + y &= x^3 - x^2 - 10x - 20 \\
A_1 : y^2 + y &= x^3 - x^2 \\
A_2 : y^2 + y &= x^3 - x^2 - 7820x - 263580.
\end{aligned}
$$

In fact, all three curves are isogenous over $\mathbb{Q}$, and there are no other elliptic curves of conductor 11 defined over $\mathbb{Q}$ (cf. [Cr] and [Wi]). The curves have split multiplicative reduction at 11, good ordinary reduction at a set of primes of density 1, commencing with $3, 5, 7, 13, 17, \ldots$, and good supersingular reduction at an infinite set of primes starting with $2, 19, \ldots$. In fact, Conjecture 2.5 is true for all primes p for all the three curves A_i over $\mathbb{Q}$, $(i = 0, 1, 2)$, by the following arguments.

It is well-known [Cr] that the complex L-function $L(A_i/\mathbb{Q}, s)$ of all three curves A_i does not vanish at $s = 1$. Hence, by [C-M], the dual of $H^1(G_{S,\infty}, A_{i,p^\infty})$ has $\Lambda(\Gamma)$-rank equal to 1 for all primes p. The case $p = 2$ is not dealt with in [C-M], but similar arguments work in this case too. By (8), it follows that $\mathcal{S}(\widehat{E/K_\infty})$ has $\Lambda(\Gamma)$-rank at most 1 for all primes p. Conjecture 2.5 is then clear for all primes where the A_i have good supersingular reduction from Theorem 2.6. The proof of Conjecture 2.5 for primes of good ordinary reduction follows from Theorem 2.8 below, since the deep theorem of Kolyvagin shows that in this case $\mathcal{S}(A_i/\mathbb{Q})$ is finite for all primes p because $L(A_i/\mathbb{Q}, 1) \neq 0$, for $i = 0, 1, 2$. A similar argument also works for $p = 11$, since a simple numerical argument proves that the 11-adic logarithm of the 11-adic Tate period of the curves A_i over $\mathbb{Q}_{11}$ is not zero. These curves will be treated more fully in Chapters 4 and 5.

It is one of the miracles of cyclotomic theory that there is an easy proof of an important case of Conjecture 2.5.

2.8. Theorem. *Assume that the prime p satisfies the following:*

(i) *E has potential good ordinary reduction at all primes v of F dividing p,*

(ii) *$\mathcal{S}(E/F)$ is finite (or equivalently both $E(F)$ and $\text{III}(E/F)(p)$ are finite).*

Then $\widehat{\mathcal{S}(E/K_\infty)}$ is $\Lambda(\Gamma)$-torsion.

An important corollary of Theorem 2.8 is the following.

2.9. Corollary. *Let E be a modular elliptic curve over $\mathbb{Q}$ with the property that its complex L-function $L(E/\mathbb{Q}, s)$ does not vanish at $s = 1$. Then $\widehat{\mathcal{S}(E/K_\infty)}$ is $\Lambda(\Gamma)$-torsion for all primes p where E has potential ordinary reduction.*

Proof. By a deep theorem of Kolyvagin [Ko], under the above hypotheses, the group $\mathcal{S}(E/\mathbb{Q})$ is finite for every prime p. The corollary now follows from Theorem 2.8.

$\square$

The following classical lemma from the theory of $\mathbb{Z}_p$-extensions is needed in the proof of Theorem 2.8.

2.10. Lemma. *Let X be a finitely generated $\Lambda(\Gamma)$-module. If X_Γ is finite, then X is a torsion $\Lambda(\Gamma)$-module.*

We stress that the proof of this lemma (cf. [W], see Appendix, Corollary A.1.6), depends crucially on the structure theory of finitely generated $\Lambda(\Gamma)$-modules. It is the only place in this chapter where we use the structure theory of $\Lambda(\Gamma)$-modules. If we take $H_\infty = F(E_{p^\infty})$, where E is assumed to have no complex multiplication, then the Galois group $\Omega = G(H_\infty/F)$ is an open subgroup of $GL_2(\mathbb{Z}_p)$. It is shown in [B-H] that the lemma is false in this case for the corresponding Iwasawa algebra. As Susan Howson has spointed out, the correct analogue of

Lemma 2.10 in this case involves the $\mathbb{Z}_p$-coranks of all the higher cohomology groups pf Ω acting on the Pontryagin dual of X (not just the H^0, as in Lemma 2.10).

2.11. Proof of Theorem 2.8. We apply Lemma 2.10 with $X = \widehat{\mathcal{S}(E/K_\infty)}$, so that X_Γ is dual to $B = \mathcal{S}(E/K_\infty)^\Gamma$. Hence we must show that the conditions imposed on p in Theorem 2.8 imply that B is finite. To do this, we appeal to the fundamental diagram (9). We recall that the map β in (9) is surjective. Thus, by the snake lemma and the finiteness of $\mathcal{S}(E/F)$, B will certainly be finite if we can show that Ker γ is finite. If $v \in S$ and v does not divide p, then Ker γ_v is clearly finite by Lemma 1.11. If v divides p, we will now explain how the finiteness of Ker γ_v follows from the results of [C-G] - and of course, it is here that we will use our hypothesis that E has potential ordinary reduction at all primes v dividing p.

Fix a prime v of F dividing p, and let Γ_v denote the decomposition group of v in Γ. We also write v for some fixed prime of K_∞ that extends the prime v of F. We have a tower of extensions:

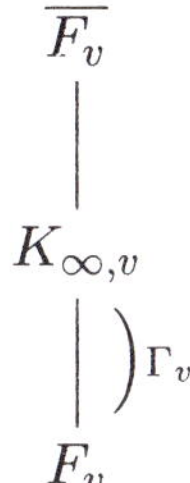

$$\overline{F_v}$$
$$|$$
$$K_{\infty,v}$$
$$\left.\middle|\;\right)\Gamma_v$$
$$F_v$$

By the inflation-restriction sequence,

$$\text{Ker}\ \gamma_v = H^1(\Gamma_v, E(K_{\infty,v})).$$

As is explained in [C-G, p.130] (see Appendix A.2), Tate duality implies that the group Ker γ_v is dual to $E(F_v)/E_U(K_{\infty,v})$, where $E_U(K_{\infty,v})$ is the group of universal norms in $E(F_v)$ for the $\mathbb{Z}_p$-extension $K_{\infty,v}/F_v$, i.e.,

$$E_U(K_{\infty,v}) = \bigcap_{K'} N_{K'/F_v}(E(K'))$$

where K' runs over all finite extensions of F_v contained in $K_{\infty,v}$ and N_{K'/F_v} denotes the norm map from K' to F_v on E.

We now show, using the results of [C-G] (see Appendix A.2), that $E_U(K_{\infty,v})$ is of finite index in $E(F_v)$. By hypothesis, there exists a finite extension L_v of F_v such that E/L_v has good ordinary reduction. Then, in the terminology of [C-G], the composite extension $L_{\infty,v} = L_v K_{\infty,v}$ is a deeply ramified p-adic field because it is a ramified $\mathbb{Z}_p$-extension of F_v (we recall that each prime of F dividing p is ramified in the cyclotomic $\mathbb{Z}_p$-extension K_∞, cf. [W]). Hence by Propositions 5.5 and 5.6 of [C-G] (see Appendix A.2.5), the group $E_U(L_{\infty,v})$ of universal norms in $E(L_v)$ for the $\mathbb{Z}_p$-extension $L_{\infty,v}/L_v$ has finite index. We have:

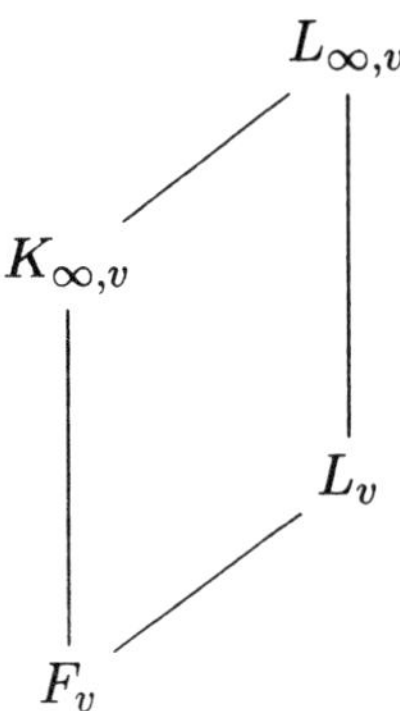

But clearly,

$$N_{L_v/F_v}(E_U(L_{\infty,v})) \subseteq E_U(K_{\infty,v}) \subseteq E(F_v)$$

as the norm map sends any open subgroup of $E(L_v)$ onto an open subgroup of $E(F_v)$, it follows that $E_U(K_{\infty,v})$ certainly has finite index in $E(F_v)$. Hence Ker γ_v is finite for all v dividing p, and the proof of Theorem 2.8 is complete.

$\square$

2.12. Remark. There are various ways of deducing that $E_U(L_{\infty,v})$ has finite index in $E(L_v)$, or equivalently that $H^1(G(L_{\infty,v}/L_v), E(L_{\infty,v}))$ is finite when E has good ordinary reduction over L by using the main results of [C-G]. A second argument to do this is given in the proof of Proposition 3.5 in Chapter 3.

2.13. Proof of Theorem 2.6. By the result of Greenberg [G, Proposition 3], the dual of $H^1(G_{S,\infty}, E_{p^\infty})$ has $\Lambda(\Gamma)$-rank $\geq [F:\mathbb{Q}]$. On the other hand, we make the following claim:

$$\Lambda(\Gamma)\text{-rank of } \bigoplus_{v \in S} \widehat{J_v(K_\infty)} = [F:\mathbb{Q}] - r_p(E/F). \tag{11}$$

We remark that the proof of Theorem 2.6 is immediate from these estimates and the exact sequence (8).

We proceed to prove the claim above using the local arguments in [C-G]. Let v be any place in S, and let $\Gamma_v \subseteq \Gamma$ be the decomposition group of v in $\Gamma = G(K_\infty/F)$. Since

$$J_v(K_\infty) = \bigoplus_{w|v} H^1(K_{\infty,w}, E)(p),$$

and the number of summands in the direct sum on the right is $[\Gamma : \Gamma_v]$, it is clear that the $\Lambda(\Gamma)$-rank of $J_v(K_\infty)$ is equal to t_w, where

$$t_w = \Lambda(\Gamma_v)\text{-rank of } H^1(\widehat{K_{\infty,w}}, E)(p),$$

and w is some fixed prime of K_∞ above v. Also, Kummer theory (take direct limits in the Kummer sequence , cf. (**1.1**)) gives the exact sequence

$$0 \longrightarrow E(K_{\infty,w})\otimes \mathbb{Q}_p/\mathbb{Z}_p \longrightarrow H^1(K_{\infty,w}, E_{p^\infty}) \longrightarrow H^1(K_{\infty,w}, E)(p) \longrightarrow 0.$$

Hence we have

$$t_w \leq \Lambda(\Gamma_v) - \text{rank of} H^1(\widehat{K_{\infty,w}}, E_{p^\infty}).$$

If v does not divide p, then [G, Proposition 2] shows that $H^1(\widehat{K_{\infty,w}}, E_{p^\infty})$ is $\Lambda(\Gamma_v)$-torsion and therefore $t_w = 0$. On the other hand, if v does divide p, we need the finer bounds for t_w obtained by using [C-G, Proposition 4.9]. In this case, the integer g occuring in *loc. cit.* is 1, while the integer h occuring there has the value 1 or 2 accordingly as $v \notin P_p(E/F)$ or $v \in P_p(E/F)$ (cf. [C-G, p.150]). This is because $v \in P_p(E/F)$ if and only if the Galois submodule C of E_{p^∞} defined in [C-G, p.150] is equal to the whole of E_{p^∞}. By [C-G, Proposition 4.9], we therefore have

$$t_w = \begin{cases} [F_v : \mathbb{Q}_p] & \text{if } v \notin P_p(E/F) \\ 0 & \text{if } v \in P_p(E/F). \end{cases}$$

This proves (11) since $[F : \mathbb{Q}] = \sum_{v|p}[F_v : \mathbb{Q}_p]$ and the proof of Theorem 2.6 is now complete.

$\square$

Theorem 2.8 has the following simple generalization, giving further examples where Conjecture 2.5 can be proven.

2.14. Theorem. *Assume that p satisfies the following:*

(i) *E has potential good reduction at all primes v of F dividing p.*

(ii) *$\mathcal{S}(E/F)$ is finite.*

Then $\widehat{\mathcal{S}(E/K_\infty)}$ has $\Lambda(\Gamma)$-rank equal to $r_p(E/F)$.

Proof. The crucial point of the proof is to show that the $\mathbb{Z}_p$-corank of $\mathcal{S}(E/K_\infty)^\Gamma$ is exactly equal to $r_p(E/F)$. Indeed, granted the assertion, we let t be the $\Lambda(\Gamma)$-rank of $\widehat{\mathcal{S}(E/K_\infty)}$. Then it is clear from the structure theory of finitely generated $\Lambda(\Gamma)$-modules that $\widehat{\mathcal{S}(E/K_\infty)}_\Gamma$ has $\mathbb{Z}_p$-rank at least t. But $\widehat{\mathcal{S}(E/K_\infty)}_\Gamma$ is dual to $\mathcal{S}(E/K_\infty)^\Gamma$, and also $t \geq r_p(E/F)$ by Theorem 2.6. Hence the above assertion shows that necessarily $t = r_p(E/F)$, as required.

We now proceed to show that $\mathcal{S}(E/K_\infty)^\Gamma$ has $\mathbb{Z}_p$-corank equal to $r_p(E/F)$ by analysing the fundamental diagram (9). We recall that β is surjective since Γ has p-cohomological dimension equal to 1. Also, Ker β is finite by Imai's theorem [I] (see also the Appendix). Our hypothesis that $\mathcal{S}(E/F)$ is finite shows that Im α is finite. Also, by Proposition 1.9, $\mathrm{Coker}(\lambda_S(F))$ is finite. Hence we see on applying the snake lemma to (9) that our assertion on the $\mathbb{Z}_p$-corank of $\mathcal{S}(E/K_\infty)^\Gamma$ is equivalent to the statement that Ker γ has $\mathbb{Z}_p$-corank equal to $r_p(E/F)$. But γ is the direct sum of the local maps γ_v for v in S. By Lemma 1.11, Ker γ_v is finite for v not dividing p. As above, put $t_v = 0$ or $[F_v : \mathbb{Q}_p]$, according as E has potential ordinary or potential supersingular reduction at v. Thus $r_p(E/F) = \sum_{v|p} t_v$. We claim that $\mathbb{Z}_p$-corank of Ker γ_v is equal to t_v. If E has potential ordinary reduction at v, this is simply the assertion that Ker γ_v is finite, which is established in the proof of Theorem 2.8. If E has potential supersingular reduction at v, then the results of [C-G] show that γ_v is the zero map (cf. [C-G, Proposition 4.9]), and hence Ker $\gamma_v = H^1(F_v, E)(p)$. Then, by Tate duality, $H^1(F_v, E)(p)$ is dual to $E(F_v)^*$, and this latter group has $\mathbb{Z}_p$-rank t_v. This completes the proof of Theorem 2.14.

□

THE DIVISION FIELD CASE

2.15. For the whole of this section, we assume that E does not admit complex multiplication, i.e., that $\mathrm{End}_{\overline{F}}(E) = \mathbb{Z}$. We let $F_\infty = F(E_{p^\infty})$, and we write $\Sigma = G(F_\infty/F)$ for the Galois group of F_∞ over F. The action of Σ on E_{p^∞} defines an injection of Σ into $\text{Aut}(E_{p^\infty}) = GL_2(\mathbb{Z}_p)$, and a fundamental theorem of Serre [Se2] asserts that the image of Σ is open in $GL_2(\mathbb{Z}_p)$ for all primes p (and equal to $GL_2(\mathbb{Z}_p)$ for all but a finite number of primes p). Thus the Iwasawa algebra (cf. [H]) $\Lambda(\Sigma)$ is non-abelian, and very little is known about its representation theory, in particular. Nevertheless, many of the results which we have just discussed in the cyclotomic case have natural generalizations to this non-abelian situation, which was first studied in [H1]. We briefly state, without proof, what is known at present, referring the interested reader to [C], [H], [C-H].

Again, it is true that the dual of the Selmer group, $\mathcal{S}(\widehat{E/F_\infty})$, is finitely generated over the Iwasawa algebra $\Lambda(\Sigma)$. We can define the $\Lambda(\Sigma)$-rank of any finitely generated $\Lambda(\Sigma)$-module as follows. By a theorem of Lazard [La] we can find an open subgroup Φ of Σ such that the sub-algebra $\Lambda(\Phi)$ is noetherian and has no zero divisors. It is then well-known that $\Lambda(\Phi)$ has a skew field of fractions, which we denote by $Q(\Phi)$, and the $\Lambda(\Phi)$-rank of X is then defined to be the dimension of the $Q(\Phi)$-vector space $Q(\Phi) \otimes_{\Lambda(\Phi)} X$. It is convenient to remove the dependence on the subgroup Φ by defining the $\Lambda(\Sigma)$-rank of X to be the $\Lambda(\Phi)$-rank of X divided by the index $[\Sigma : \Phi]$. Note that the $\Lambda(\Sigma)$-rank of X is not always an integer.

2.16. Conjecture. *For every prime p, the $\Lambda(\Sigma)$-rank of $\mathcal{S}(\widehat{E/F_\infty})$ is equal to $r_p(E/F)$, where $r_p(E/F)$ is defined by (10).*

One can prove an exact analogue of Theorem 2.6 and a slightly stronger form of it over F_∞ (see [C] or [H])

2.17. Theorem. *For all primes p, we have*

$$r_p(E/F) \leq \Lambda(\Sigma)\text{-rank of } \mathcal{S}(\widehat{E/F_\infty}) \leq [F : \mathbb{Q}].$$

2.18. When E has potential supersingular reduction at all primes v of F dividing p, Theorem 2.17 proves Conjecture 2.16. However, very little is known at present about Conjecture 2.16 when $r_p(E/F) < [F : \mathbb{Q}]$, and in particular, about Conjecture 2.16 when E has potential ordinary reduction at all primes v of F dividing p. In fact, in this latter case, Conjecture 2.16 has so far only been proven for a small number of isogeny classes of elliptic curves over $\mathbb{Q}$ and in which $p = 5$ or $p = 7$, including the isogeny class of the three elliptic curves A_0, A_1, A_2 of conductor 11, and the prime $p = 5$ (see [C] or [C-H]). The arguments used to prove Conjecture 2.16 in these examples were inspired by the interesting paper [H-M]. It is also striking that all of these numerical examples have the following features in common:

(i) For all E in the isogeny class, $\mathbb{Q}(E_{p^\infty})$ is a pro-p-extension of $\mathbb{Q}(\mu_p)$.

(ii) There exists an E in the isogeny class such that $\mathcal{S}(E/\mathbb{Q}(\mu_{p^\infty})) = 0$.

We shall give a detailed proof of (i) and (ii) for the isogeny class A_0, A_1, A_2 of curves of conductor 11 and $p = 5$ in Chapter 5.

Let us also mention the following basic difference between the cyclotomic case and the division field case (see [C-H]), which was first pointed out to us by R. Greenberg.

2.19. Theorem. *For every prime $p \geq 5$, the $\mathbb{Q}_p$-vector space*

$$\widehat{\mathcal{S}(E/F_\infty)} \otimes_{\mathbb{Z}_p} \mathbb{Q}_p$$

is infinite dimensional.

At present, we have no idea when this large Selmer group over F_∞ is due to the existence of many independent points of infinite order which are rational over F_∞, or to the existence of a Tate-Shafarevich group of E/F_∞ with a large p-primary component. For example, it is unknown whether or not any of the three curves $E = A_i$, $0 \leq i \leq 2$ of conductor 11 have a single point of infinite order over the field $\mathbb{Q}(E_{5^\infty})$ of 5-power division points. We show in Chapter 5 that $E = A_i$, $0 \leq i \leq 2$ has no points of infinite order in the field $Q(\mu_{5^\infty})$.

Chapter 3

The Euler Characteristic Formula

INTRODUCTION

3.1. Again E will be an elliptic curve defined over a finite extension F of $\mathbb{Q}$, and p will be an arbitrary prime number. Moreover, as in Chapter 2, H_∞ will denote an infinite Galois extension of F, whose Galois group $\Omega = G(H_\infty/F)$ is a p-adic Lie group of positive dimension. Let $\mathcal{S}(E/H_\infty)$ be the Selmer group of E over H_∞. If this Selmer group is to be useful in studying the arithmetic of E over the base field F, we must be able to recover the basic arithmetic invariants of E over F from some exact formula attached to $\mathcal{S}(E/H_\infty)$. The simplest means of obtaining such an exact formula is to calculate the Ω-Euler characteristic of $\mathcal{S}(E/H_\infty)$. This is the question we shall study in this chapter, concentrating mainly on the case when H_∞ is the cyclotomic $\mathbb{Z}_p$-extension of F. In all cases in which we have been able to calculate the Ω-Euler characteristic of $\mathcal{S}(E/H_\infty)$, it turns out to be very closely related to the conjectural exact formula of Birch and Swinnerton-Dyer for the leading term in the development of the complex L-function of E over F at the point $s = 1$ in the complex plane. Although we shall not discuss the aspects of analytic Iwasawa theory further in these notes, this provides a method for attacking the conjecture of Birch and Swinnerton-Dyer whenever we can prove a "main conjecture" for the $\Lambda(\Omega)$-module $\mathcal{S}(E/H_\infty)$.

To calculate Euler characteristics, we must study all of the cohomology groups $H^i(\Omega, \mathcal{S}(E/H_\infty))$ $(i = 0, 1, \dots)$ and it is useful to make the following general remarks. Let d be the dimension of Ω as a p-adic Lie group. Then by Serre's refinement [Se3] of Lazard's theorem, Ω will have p-cohomological dimension equal to d provided Ω has no non-trivial p-torsion. Moreover, the fundamental diagram (9) relates the finiteness of $H^0(\Omega, \mathcal{S}(E/H_\infty))$ to the following issues. Let us assume that both Ker β

29

and Coker β are finite (this is true when H_∞ is either the cyclotomic $\mathbb{Z}_p$-extension of F or the field $F(E_{p^\infty})$). Then it follows from (9) that $H^0(\Omega, \mathcal{S}(E/H_\infty))$ is finite if and only if both $\mathcal{S}(E/F)$ and $\mathrm{Ker}\ \gamma \cap \mathrm{Im}\ \lambda$ are finite. In practice, we shall assume that $\mathcal{S}(E/F)$ is finite and use the local methods of [C-G] to show that $\mathrm{Ker}\ \gamma$ is finite under suitable conditions on the extension H_∞ and the nature of the reduction of E at the primes v of F dividing p.

CYCLOTOMIC THEORY

3.2. We now take H_∞ to be the cyclotomic $\mathbb{Z}_p$-extension K_∞ of F, and as in Chapter 2, we write $\Gamma = G(K_\infty/F)$. Since Γ is topologically isomorphic to $\mathbb{Z}_p$, it has p-cohomological dimension equal to 1. Our aim is to calculate the Γ-Euler characteristic of $\mathcal{S}(E/K_\infty)$ under appropriate conditions on E and p. In general, if A is a discrete p-primary Γ-module, we shall say that A has finite Γ-Euler characteristic if both $H^0(\Gamma, A)$ and $H^1(\Gamma, A)$ are finite, and we then define its Euler characteristic by

$$\chi(\Gamma, A) = \#\ H^0(\Gamma, A)/\#\ H^1(\Gamma, A). \tag{12}$$

For each finite place v of F, let $c_v = [E(F_v) : E_0(F_v)]$, where, as in Chapter 1, $E_0(F_v)$ denotes the subgroup of points with non-singular reduction. We write k_v for the residue field at v, and $\tilde{E}_v$ over k_v for the reduction of E modulo v. Assuming $\mathcal{S}(E/F)$ is finite, we define

$$\rho_p(E/F) = \#\mathrm{III}(E/F)(p)/\#(E(F)(p))^2 \times \prod_v c_v^{(p)} \times \prod_{v|p} (d_v^{(p)})^2, \tag{13}$$

where $d_v = \#(\tilde{E}_v(k_v))$, and for any positive integer n, $n^{(p)}$ denotes the largest power of p dividing n. The following is the main result proven in this chapter.

3.3. Theorem. (Main Theorem) *Assume that*

(i) *$p > 2$,*

(ii) *E has good ordinary reduction at all places v of F dividing p,*

(iii) *$\mathcal{S}(E/F)$ is finite.*

Then $\mathcal{S}(E/K_\infty)$ has finite Γ-Euler characteristic given by

$$\chi(\Gamma, \mathcal{S}(E/K_\infty)) = \rho_p(E/F).$$

The proof of the theorem is rather involved. We immediately give the proof of this result and postpone the discussion of numerical examples until the next chapters. The essence of the proof is to make a more detailed study of the fundamental diagram (9). We start by analyzing the map $\gamma = \underset{v \in S}{\oplus}\, \gamma_v$, which is a purely local calculation that does not need the hypothesis that $\mathcal{S}(E/F)$ is finite. Recall that for any $v \in S$,

$$J_v(K_\infty) = \underset{w|v}{\oplus}\, H^1(K_{\infty,w}, E)(p).$$

Let Γ_v denote the decomposition group of v in Γ. By Shapiro's lemma, we have

$$H^i(\Gamma, J_v(K_\infty)) \simeq H^i(\Gamma_v, H^1(K_{\infty,w}, E)(p)), \qquad \forall\, i \geq 0$$

for some fixed prime w of K_∞ above v. Hence by the Hochschild-Serre spectral sequence, and the fact that $H^2(F_v, E) = 0$ [Lemma 1.12], we obtain

$$\mathrm{Ker}\ \gamma_v = H^1(\Gamma_v, E(K_{\infty,w})), \quad \mathrm{Coker}\ \gamma_v = H^2(\Gamma_v, E(K_{\infty,w})).$$

(Note that these groups are p-primary since Γ_v is a pro-p group).

3.4. Lemma. *Assume that v does not divide p. Then γ_v is surjective, and* $\mathrm{Ker}\ \gamma_v$ *is finite of order* $c_v^{(p)}$.

Proof. Let F_v^{nr} be the maximal unramified extension of F_v, and put

$$W_v = H^1(G(F_v^{nr}/F_v), E(F_v^{nr})).$$

It is well-known (**1.1**) that W_v is the exact orthogonal complement of $E_0(F_v)$ under the dual Tate pairing of $H^1(F_v, E)$ and $E(F_v)$. Hence W_v is dual to $E(F_v)/E_0(F_v)$ and

$$\#\, W_v(p) = c_v^{(p)}.$$

Now assume that v does not divide p. Then $K_{\infty,w}$ is the unique unramified $\mathbb{Z}_p$-extension of F_v, whence $\mathrm{Ker}\ \gamma_v = W_v(p)$, and so the proof of the second part of the lemma is complete. As for the first part, consider the following commutative diagram, where the vertical maps are the restriction maps and the horizontal ones are the natural maps arising from Kummer theory:

$$
\begin{array}{ccc}
H^1(F_v, E_{p^\infty}) & \longrightarrow & H^1(F_v, E)(p) \\
{\scriptstyle \gamma_v'}\downarrow & & \downarrow{\scriptstyle \gamma_v} \\
H^1(K_{\infty,w}, E_{p^\infty})^{\Gamma_v} & \longrightarrow & H^1(K_{\infty,w}, E)(p)^{\Gamma_v}.
\end{array}
\tag{14}
$$

We claim that the horizontal maps are in fact isomorphisms. Indeed, Kummer theory **(1.1)** shows that the maps are surjective, and they are injective because

$$E(F_v) \otimes \mathbb{Q}_p/\mathbb{Z}_p = 0 \qquad \text{and} \qquad E(K_{\infty,w}) \otimes \mathbb{Q}_p/\mathbb{Z}_p = 0,$$

as v does not divide p. The restriction map γ_v' is surjective since the cokernel maps to $H^2(\Gamma_v, E_{p^\infty})$ and Γ_v has p-cohomological dimension 1. The surjectivity of γ_v now follows from (14).

$\square$

3.5. Proposition. *Assume that v divides p and that E has good ordinary reduction at v. Then γ_v is surjective, and* Ker γ_v *is finite of order* $(d_v^{(p)})^2$; *where* $d_v = \#(\tilde{E}_v(k_v))$.

Proof. Let $\mathfrak{M}_{\infty,w}$ be the maximal ideal of the ring of integers of $K_{\infty,w}$, and let $k_{\infty,w}$ be the residue field. Let $\hat{E}_v$ be the formal group over the ring of integers of F_v which gives the kernel of reduction modulo v on E. Then we have the exact sequence

$$0 \longrightarrow \hat{E}_v(\mathfrak{M}_{\infty,w}) \longrightarrow E(K_{\infty,w}) \longrightarrow \tilde{E}_v(k_{\infty,w}) \longrightarrow 0, \tag{15}$$

where $\tilde{E}_v(k_{\infty,w})$ is a finite group since $k_{\infty,w}$ is a finite field, and $\hat{E}_v(\mathfrak{M}_{\infty,w})$ is a $\mathbb{Z}_p$-module since $\hat{E}_v$ is a formal group [Si, Chapter IV]. Note that the map $E(F_v) \to \tilde{E}_v(k_v)$ is surjective because E has good reduction at v [Si, Chapter VII]. Thus, taking Γ_v-cohomology of the exact sequence (15), we obtain the long exact sequence

$$0 \longrightarrow H^1(\Gamma_v, \hat{E}_v(\mathfrak{M}_{\infty,w})) \longrightarrow H^1(\Gamma_v, E(K_{\infty,w})) \tag{16}$$

$$\longrightarrow H^1(\Gamma_v, \tilde{E}_v(k_{\infty,w})) \longrightarrow H^2(\Gamma_v, \hat{E}_v(\mathfrak{M}_{\infty,w}))$$

$$\longrightarrow H^2(\Gamma_v, E(K_{\infty,w})) \longrightarrow 0.$$

The last zero occurs in this sequence because

$$H^2(\Gamma_v, \tilde{E}_v(k_{\infty,w})) = H^2(\Gamma_v, \tilde{E}_v(k_{\infty,w})(p)) = 0,$$

as Γ_v has p-cohomological dimension 1.

We now use the main result of [C-G] (see Appendix A.2) for the field $K_{\infty,w}$ and the formal group $\hat{E}_v$. Here we note that $K_{\infty,w}$ is indeed deeply ramified in the sense of [C-G] (see Appendix A.2.2), since it contains the cyclotomic $\mathbb{Z}_p$-extension of $\mathbb{Q}_p$. Hence by [C-G, Corollary 3.2] (see Appendix, Theorem A.2.3)

$$H^i(K_{\infty,w}, \hat{E}_v(\overline{\mathfrak{M}})) = 0 \text{ for } i \geq 1, \tag{17}$$

where $\overline{\mathfrak{M}}$ denotes the maximal ideal of the ring of integers of an algebraic closure of F_v.

$$
\begin{array}{c}
\overline{F_v} \\
| \\
| \\
K_{\infty,w} \\
\Big| \Big) \Gamma_v \\
F_v
\end{array}
$$

By the inflation-restriction sequence [Se], it follows that we have the exact sequence

$$0 \longrightarrow H^i(\Gamma_v, \hat{E}_v(\mathfrak{M}_{\infty,w})) \longrightarrow H^i(F_v, \hat{E}_v(\overline{\mathfrak{M}})) \longrightarrow H^i(K_{\infty,w}, \hat{E}_v(\overline{\mathfrak{M}}))^{\Gamma_v}$$

for all $i \geq 1$. Hence by (17), we obtain isomorphisms

$$H^i(\Gamma_v, \hat{E}_v(\mathfrak{M}_{\infty,w})) \simeq H^i(F_v, \hat{E}_v(\overline{\mathfrak{M}})) \text{ for } i \geq 1. \tag{18}$$

We now show that Proposition 3.5 follows from the lemma below, whose proof will be given shortly.

3.6. Lemma. *Under the hypotheses of Proposition 3.5, we have*

(i) *$H^1(F_v, \hat{E}_v(\overline{\mathfrak{M}}))$ is finite of order $d_v^{(p)} = \# \, \tilde{E}_v(k_v)(p)$,*

(ii) *$H^i(F_v, \hat{E}_v(\overline{\mathfrak{M}})) = 0$ for all $i \geq 2$.*

Indeed, by Lemma 3.6 (ii) and (18), we see that $H^2(\Gamma_v, \hat{E}_v(\mathfrak{M}_{\infty,w})) = 0$, and so by (16) it follows that

$$\text{Coker } \gamma_v = H^2(\Gamma_v, E(K_{\infty,w})) = 0.$$

Moreover (16) then reduces to the short exact sequence

$$0 \longrightarrow H^1(\Gamma_v, \hat{E}_v(\mathfrak{M}_{\infty,w})) \longrightarrow \text{Ker } \gamma_v \longrightarrow H^1(\Gamma_v, \tilde{E}_v(k_{\infty,w})) \longrightarrow 0. \tag{19}$$

By Lemma 3.6 (i), the group on the left side of (19) has order $d_v^{(p)}$. On the other hand, to compute the order of the group on the right side of (19), we may replace $\tilde{E}_v(k_{\infty,w})$ by its p-primary subgroup since Γ_v is pro-p. Further, as Γ_v is isomorphic to $\mathbb{Z}_p$ and $\tilde{E}_v(k_{\infty,w})$ is finite, we note that

$$\# \, H^1(\Gamma_v, \tilde{E}_v(k_{\infty,w})(p)) \;=\; \# \, H^0(\Gamma_v, \tilde{E}_v(k_{\infty,w})(p)) \;=\; d_v^{(p)}.$$

Hence (19) shows that Ker γ_v has order $(d_v^{(p)})^2$, as required. Thus the proof of Proposition 3.5 is complete once we prove Lemma 3.6.

3.7. Proof of Lemma 3.6. Consider the Weil pairing

$$E_{v,p^n} \times E_{v,p^n} \to \mu_{p^n}.$$

It is well-known that under the Weil pairing, $\hat{E}_{v,p^n}$ is its own orthogonal complement and therefore

$$\tilde{E}_{v,p^n} = \mathrm{Hom}(\hat{E}_{v,p^n}, \mu_{p^n}) \text{ for } n \geq 1. \tag{20}$$

Since $\# \, \hat{E}_{v,p^n}(\overline{F_v}) = p^n$, by Tate's local Euler characteristic theorem (cf. **(1.1)**), we have

$$\# \, H^1(F_v, \hat{E}_{v,p^n}) = p^{dn+a+b},$$

where

$$d = [F_v : \mathbb{Q}_p], \;\; p^a = \# \, H^0(F_v, \hat{E}_{v,p^n}), \;\; p^b = \# \, H^2(F_v, \hat{E}_{v,p^n}).$$

Assume for the rest of the proof that $n >> 0$. Now, by Tate local duality **(1.1)** and (20), we have

$$p^b = \# \, H^2(F_v, \hat{E}_{v,p^n}) = \# \, H^0(F_v, \tilde{E}_{v,p^n}) = \#(\tilde{E}_v(k_v)_{p^n}) = d_v^{(p)}. \tag{21}$$

Consider the exact sequence

$$0 \longrightarrow \hat{E}_{v,p^n} \longrightarrow \hat{E}_{v,p^\infty} \xrightarrow{p^n} \hat{E}_{v,p^\infty} \longrightarrow 0$$

of $G(\overline{F_v}/F_v)$-modules. By our assumption on n, the associated long exact Galois cohomology sequence gives

$$0 \longrightarrow \hat{E}_{v,p^\infty}(F) \longrightarrow H^1(F_v, \hat{E}_{v,p^n}) \longrightarrow H^1(F_v, \hat{E}_{v,p^\infty})_{p^n} \longrightarrow 0.$$

Since $\# \hat{E}_{v,p^\infty}(F) = p^a$, we have

$$\# H^1(F_v, \hat{E}_{v,p^\infty})_{p^n} = p^{dn+b}. \tag{22}$$

Therefore

$$H^1(F_v, \hat{E}_{v,p^\infty}) \simeq (\mathbb{Q}_p/\mathbb{Z}_p)^d \oplus B \tag{23}$$

where B is a finite group with $\# B = p^b = d_v^{(p)}$ by (21).

We also have the exact sequence

$$0 \longrightarrow \hat{E}_{v,p^n} \longrightarrow \hat{E}_v(\overline{\mathfrak{M}}) \overset{p^n}{\longrightarrow} \hat{E}_v(\overline{\mathfrak{M}}) \longrightarrow 0. \tag{24}$$

Taking $G(\overline{F_v}/F_v)$-cohomology and then taking the direct limit as $n \to \infty$, we obtain the exact sequence

$$0 \longrightarrow \hat{E}_v(\mathfrak{M}) \otimes \mathbb{Q}_p/\mathbb{Z}_p \longrightarrow H^1(F_v, \hat{E}_{v,p^\infty}) \longrightarrow H^1(F_v, \hat{E}_v(\overline{\mathfrak{M}})) \longrightarrow 0.$$

Since the group on the left of this sequence is divisible of $\mathbb{Z}_p$-corank d, it must be the maximal divisible subgroup of $H^1(F, \hat{E}_{v,p^\infty})$ by (23), and so it follows from (23) that

$$H^1(F_v, \hat{E}_v(\overline{\mathfrak{M}})) = \# B = d_v^{(p)}.$$

This proves (i) in the lemma.

To prove (ii), we take the $G(\overline{F_v}/F_v)$-cohomology of (24) and obtain the exact sequence

$$H^1(F_v, \hat{E}_v(\overline{\mathfrak{M}})) \overset{p^n}{\longrightarrow} H^1(F_v, \hat{E}_v(\overline{\mathfrak{M}})) \overset{\alpha}{\longrightarrow} H^2(F_v, \hat{E}_{v,p^n})$$

$$\longrightarrow (H^2(F_v, \hat{E}_v(\overline{\mathfrak{M}})))_{p^n} \longrightarrow 0.$$

By (i), p^n annihilates $H^1(F_v, \hat{E}_v(\overline{\mathfrak{M}}))$ (recall that we have chosen $n >> 0$), and so α is injective. But then α is surjective by (i) and (21). Hence $H^2(F_v, \hat{E}_v(\overline{\mathfrak{M}})) = 0$. This completes the proof of Lemma 3.6.

$\square$

The following lemma, first established by Imai [I], can be proven by local or global means (see the Appendix Theorem A.2.8 for an outline of one proof).

3.8. Lemma. *The group $H^0(K_\infty, E_{p^\infty})$ is finite.*

We now return to the proof of Theorem 3.3. We split the fundamental diagram (9) into two commutative diagrams with exact rows, namely

$$
\begin{array}{ccccccccc}
0 & \longrightarrow & \mathcal{S}(E/K_\infty)^\Gamma & \longrightarrow & H^1(G_{S,\infty}, E_{p^\infty})^\Gamma & \longrightarrow & \text{Im } \phi_\infty & \longrightarrow & 0 \\
& & \uparrow{\scriptstyle\alpha} & & \uparrow{\scriptstyle\beta} & & \uparrow{\scriptstyle\delta} & & \\
0 & \longrightarrow & \mathcal{S}(E/F) & \longrightarrow & H^1(G_S, E_{p^\infty}) & \longrightarrow & \text{Im } \lambda & \longrightarrow & 0
\end{array}
\tag{25}
$$

and

$$
\begin{array}{ccccccccc}
0 & \longrightarrow & \text{Im } \phi_\infty & \longrightarrow & \underset{v\in S}{\oplus} J_v(K_\infty)^\Gamma & \longrightarrow & \text{Coker } \phi_\infty & \longrightarrow & 0 \\
& & \uparrow{\scriptstyle\delta} & & \uparrow{\scriptstyle\gamma} & & \uparrow{\scriptstyle\epsilon} & & \\
0 & \longrightarrow & \text{Im } \lambda & \longrightarrow & \underset{v\in S}{\oplus} J_v(F) & \longrightarrow & \text{Coker } \lambda & \longrightarrow & 0.
\end{array}
\tag{26}
$$

where δ and ϵ are the obvious induced maps. Note that Ker β, and therefore Ker α, are finite by Lemma 3.8 and the Hochschild-Serre spectral sequence. Also, β is surjective because Γ has p-cohomological dimension 1. Moreover, Ker δ is finite because it is contained in Ker γ, which is finite by Lemma 3.4 and Proposition 3.5. Recall that we are assuming that $\mathcal{S}(E/F)$ is finite, whence it has the same order as $\text{III}(E/F)(p)$ (see Chapter 1). Applying the snake lemma to (25), we obtain the exact sequence

$$
0 \longrightarrow \text{Ker } \alpha \longrightarrow \text{Ker } \beta \longrightarrow \text{Ker } \delta \longrightarrow \text{Coker } \alpha \longrightarrow 0,
$$

whence

$$
\begin{aligned}
\# \, \mathcal{S}(E/K_\infty)^\Gamma / \# \, \text{III}(E/F)(p) &= \# \, \text{Coker } \alpha / \# \, \text{Ker } \alpha \\
&= \# \, \text{Ker } \delta / \# \, \text{Ker } \beta.
\end{aligned}
\tag{27}
$$

But, as $E_{p^\infty}(K_\infty)$ is finite, we have

$$
\# \, \text{Ker } \beta = \# \, H^1(\Gamma, E_{p^\infty}(K_\infty)) = \# \, E_{p^\infty}(K_\infty)^\Gamma = \# \, E(F)(p).
\tag{28}
$$

To compute the order of Ker δ, we apply the snake lemma to (26). We recall that γ is surjective by Lemma 3.4 and Proposition 3.5 (whence also

ϵ is surjective), and that δ is surjective because β is surjective. Hence we obtain from (26) the exact sequence

$$0 \longrightarrow \operatorname{Ker} \ \delta \longrightarrow \operatorname{Ker} \ \gamma \longrightarrow \operatorname{Ker} \ \epsilon \longrightarrow 0. \qquad (29)$$

But Lemma 3.4 and Proposition 3.5 yield

$$\# \ \operatorname{Ker} \ \gamma = \prod_{v} c_v^{(p)} \times \prod_{v|p}(d_v^{(p)})^2.$$

By Proposition 1.9 and our hypothesis that $\mathcal{S}(E/F)$ is finite, we have Coker λ is finite of order

$$\# \ \operatorname{Coker} \ \lambda = \# \ E(F)(p).$$

But Coker ϕ_∞ is also then finite, since ϵ is surjective, and

$$\# \ \operatorname{Ker} \ \epsilon = \# \ E(F)(p)/\# \ \operatorname{Coker} \ \phi_\infty.$$

Thus by (29), we have

$$\# \ \operatorname{Ker} \ \delta = (\# \ \operatorname{Coker} \ \phi_\infty/\# \ E(F)(p)) \times \prod_{v} c_v^{(p)} \times \prod_{v|p}(d_v^{(p)})^2.$$

Combining this formula with (27) and (28), we finally obtain

$$\# \ \mathcal{S}(E/K_\infty)^\Gamma/\# \ \operatorname{Coker} \ \phi_\infty = \rho_p(E/F),$$

where $\rho_p(E/F)$ is given by (13). Thus, to complete the proof of Theorem 3.3, it suffices to show that $H^1(\Gamma, \mathcal{S}(E/K_\infty))$ is finite of order

$$\# \ H^1(\Gamma, \mathcal{S}(E/K_\infty)) = \# \ \operatorname{Coker} \ \phi_\infty. \qquad (30)$$

To show this, we use the following result, whose proof will be given a little later. Note that the non-trivial part of the exactness of (31) is the surjectivity of λ_∞.

3.9. Proposition. *Under the hypotheses of Theorem 3.3, we have the exact sequence*

$$0 \longrightarrow \mathcal{S}(E/K_\infty) \longrightarrow H^1(G_{S,\infty}, E_{p^\infty}) \xrightarrow{\lambda_\infty} \underset{v \in S}{\oplus} J_v(K_\infty) \longrightarrow 0. \qquad (31)$$

Assuming the exactness of (31), we take Γ-invariants and obtain the exact sequence

$$H^1(G_{S,\infty}, E_{p^\infty})^\Gamma \xrightarrow{\ \phi_\infty\ } \underset{v\in S}{\oplus} J_v(K_\infty)^\Gamma \longrightarrow H^1(\Gamma, \mathcal{S}(E/K_\infty)) \tag{32}$$
$$\downarrow$$
$$H^1(\Gamma, H^1(G_{S,\infty}, E_{p^\infty})).$$

We claim that the group on the right hand end of this sequence is zero. Indeed, by Proposition 1.9 (i), we have $H^2(G_S, E_{p^\infty}) = 0$ since $\mathcal{S}(E/F)$ is finite. As Γ has p-cohomological dimension 1, the Hochschild-Serre spectral sequence

$$H^p(\Gamma, H^q(G_{S,\infty}, E_{p^\infty})) \Longrightarrow H^n(G_S, E_{p^\infty})$$

implies easily that

$$H^1(\Gamma, H^1(G_{S,\infty}, E_{p^\infty})) = 0, \tag{33}$$

as asserted. But (30) is now clear from (32), and so the proof of Theorem 3.3 is now complete, granted Proposition 3.9. We note that there is a well-known proof of Proposition 3.9 (see [P-R]), which works more generally under the hypothesis that $\mathcal{S}(\widehat{E/K_\infty})$ is torsion over $\Lambda(\Gamma)$. However, we shall give a different and simpler proof, which crucially uses the hypothesis that $\mathcal{S}(E/F)$ is finite.

3.10. Proof of Proposition 3.9 Let $S' = S \setminus \{v \mid p\}$. We define $\mathcal{S}'(E/F)$ and $\mathcal{S}'(E/K_\infty)$ as

$$\mathcal{S}'(E/F) \ = \ \mathrm{Ker}(H^1(G_S, E_{p^\infty}) \xrightarrow{\ \lambda'\ } \underset{v\in S'}{\oplus} J_v(F))$$

$$\mathcal{S}'(E/K_\infty) \ = \ \mathrm{Ker}(H^1(G_S, E_{p^\infty}) \xrightarrow{\ \lambda'_\infty\ } \underset{v\in S'}{\oplus} J_v(K_\infty)),$$

where λ' and λ'_∞ are the obvious localization maps. The commutative triangle

$$H^1(G_S, E_{p^\infty}) \xrightarrow{\ \lambda\ } \underset{v\in S}{\oplus} J_v(F)$$

with maps λ' and π to

$$\underset{v\in S'}{\oplus} J_v(F)),$$

where π is the projection, gives the exact sequence

$$0 \longrightarrow \mathcal{S}(E/F) \longrightarrow \mathcal{S}'(E/F) \longrightarrow \underset{v|p}{\oplus} J_v(F) \tag{34}$$

$$\overset{\theta}{\longrightarrow} \operatorname{Coker} \lambda \longrightarrow \operatorname{Coker} \lambda' \longrightarrow 0.$$

Similarly, we have the anlagous exact sequence

$$0 \longrightarrow \mathcal{S}(E/K_\infty) \longrightarrow \mathcal{S}'(E/K_\infty) \longrightarrow \underset{v|p}{\oplus} J_v(K_\infty) \tag{35}$$

$$\overset{\theta_\infty}{\longrightarrow} \operatorname{Coker} \lambda_\infty \longrightarrow \operatorname{Coker} \lambda'_\infty \longrightarrow 0.$$

Our first step will be to show that $\operatorname{Coker} \lambda'_\infty = 0$. To do this, we first note that $\operatorname{Coker} \lambda' = 0$. Indeed, by Proposition 1.9 (ii), $\operatorname{Coker} \lambda$ is dual to $E(F)(p)$ because $\mathcal{S}(E/F)$ is assumed finite. But the dual map $\widehat{\theta}$ is the natural map

$$\widehat{\theta} : E(F)(p) \rightarrow \underset{v|p}{\oplus} E(F_v)^*,$$

and thus $\widehat{\theta}$ is clearly injective, whence θ is surjective, as required.

Let

$$\phi'_\infty : H^1(G_{S,\infty}, E_{p^\infty})^\Gamma \longrightarrow \underset{v \in S'}{\oplus} J_v(K_\infty)^\Gamma$$

be the map induced by λ'_∞. Since λ' is surjective, and γ_v is surjective for all $v \in S'$ by Lemma 3.4, the obvious commutative diagram

$$
\begin{array}{ccc}
H^1(G_S, E_{p^\infty}) & \overset{\lambda'}{\longrightarrow} & \underset{v \in S'}{\oplus} J_v(F) \\
\downarrow{\scriptstyle\beta} & & \downarrow{\scriptstyle\gamma' = \underset{v \in S'}{\oplus} \gamma_v} \\
H^1(G_{S,\infty}, E_{p^\infty})^\Gamma & \overset{\phi'_\infty}{\longrightarrow} & \underset{v \in S'}{\oplus} J_v(F_\infty)^\Gamma
\end{array}
$$

shows that ϕ'_∞ is surjective. Taking Γ-cohomology of the short exact sequence

$$0 \longrightarrow \mathcal{S}'(E/K_\infty) \longrightarrow H^1(G_{S,\infty}, E_{p^\infty}) \longrightarrow \operatorname{Im} \lambda'_\infty \longrightarrow 0,$$

and noting that (33) remains valid (the argument used to prove it only uses the fact that $\mathcal{S}(E/F)$ is finite), we obtain the exact sequence

$$\longrightarrow H^1(G_{S,\infty}, E_{p^\infty})^\Gamma \longrightarrow (\operatorname{Im} \lambda'_\infty)^\Gamma \longrightarrow H^1(\Gamma, \mathcal{S}'(E/K_\infty)) \longrightarrow 0. \tag{36}$$

We also obtain that $H^1(\Gamma, \operatorname{Im} \lambda'_\infty) = 0$ because Γ has p-cohomological dimension 1. Using $H^1(\Gamma, \operatorname{Im} \lambda'_\infty) = 0$ and noting that ϕ'_∞ is surjective, it follows easily that

$$(\operatorname{Coker} \lambda'_\infty)^\Gamma = 0 \quad \text{and} \quad H^1(\Gamma, \mathcal{S}'(E/K_\infty)) = 0. \tag{37}$$

Hence Coker $\lambda'_\infty = 0$ because Coker λ'_∞ is a discrete p-primary Γ-module. We also get the second assertion of (37) for free from this argument.

Our next step will be to show that $\widehat{\operatorname{Coker} \phi_\infty}$ is $\Lambda(\Gamma)$-torsion. Since $(\widehat{\operatorname{Coker} \lambda_\infty})_\Gamma$ is dual to $(\operatorname{Coker} \lambda_\infty)^\Gamma$, it suffices by Lemma 2.10 to show that $(\operatorname{Coker} \lambda_\infty)^\Gamma$ is finite. Taking Γ-invariants of the exact sequence

$$0 \longrightarrow \mathcal{S}(E/K_\infty) \longrightarrow H^1(G_{S,\infty}, E_{p^\infty}) \longrightarrow \operatorname{Im} \lambda_\infty \longrightarrow 0,$$

and using (33) again, we obtain the exact sequence

$$\longrightarrow H^1(G_{S,\infty}, E_{p^\infty})^\Gamma \longrightarrow (\operatorname{Im} \lambda_\infty)^\Gamma \longrightarrow H^1(\Gamma, \mathcal{S}(E/K_\infty)) \longrightarrow 0,$$

and also that $H^1(\Gamma, \operatorname{Im} \lambda_\infty) = 0$. Because of this latter fact, we also have the exact sequence

$$0 \longrightarrow (\operatorname{Im} \lambda_\infty)^\Gamma \longrightarrow \big(\underset{v \in S}{\oplus} J_v(K_\infty) \big)^\Gamma \longrightarrow (\operatorname{Coker} \lambda_\infty)^\Gamma \longrightarrow 0.$$

Hence we obtain the exact sequence

$$0 \longrightarrow H^1(\Gamma, \mathcal{S}(E/K_\infty)) \longrightarrow \operatorname{Coker} \phi_\infty \longrightarrow (\operatorname{Coker} \lambda_\infty)^\Gamma \longrightarrow 0.$$

But, as noted above, Coker ϕ_∞ is finite, and so $(\operatorname{Coker} \lambda_\infty)^\Gamma$ is indeed finite. To complete the proof of Proposition 3.9, we make the following

Claim. *The $\Lambda(\Gamma)$-torsion submodule of $\underset{v|p}{\oplus} \widehat{J_v(K_\infty)}$ is zero.*

Indeed, dualizing (35) and recalling that Coker $\lambda'_\infty = 0$, it follows that the torsion module $\widehat{\operatorname{Coker} \lambda_\infty}$ injects into $\underset{v|p}{\oplus} \widehat{J_v(K_\infty)}$, and so must be zero, granted the claim. Hence it remains to prove the claim and we proceed to do so below.

Proof. (of claim) By definition,

$$J_v(K_\infty) = \underset{w|v}{\oplus} H^1(K_{\infty,w}, E)(p).$$

Pick some fixed prime w of K_∞ lying above a given prime v of F dividing p, and let Γ_v denote the decomposition group of w in Γ. It clearly suffices to show that the $\Lambda(\Gamma_v)$-module $H^1(\widehat{K_{\infty,w}, E})(p)$ has no $\Lambda(\Gamma_v)$-torsion. Here we appeal to the theory of deeply ramified fields in [C-G] (see Appendix A.2), noting that $K_{\infty,w}$ is deeply ramified because v divides p and it is a ramified $\mathbb{Z}_p$-extension of F_v. Thus, by [C-G, Proposition 4.8] (see Appendix A.2.4), we have

$$H^1(K_{\infty,w}, E)(p) \simeq H^1(K_{\infty,w}, \tilde{E}_{v,p^\infty}), \qquad (38)$$

where $\tilde{E}_{v,p^\infty}$ denotes the Galois module of all p-power torsion points on the reduction $\tilde{E}_v$ of E modulo v. Now we apply a result of Greenberg ([G, Corollary 1], see Appendix) to the right hand side of (38). Note that

$$\hat{E}_{v,p^\infty} \simeq \mathrm{Hom}(T_p(\tilde{E}_{v,p^\infty}), \mu_{p^\infty}),$$

and by a basic result of Imai ([I], see Appendix, Theorem A.2.8), we have that $\hat{E}_{v,p^\infty}(K_\infty)$ is indeed finite. Thus [G, Corollary 1] shows that the dual of the right hand side of (38) has no $\Lambda(\Gamma_v)$-torsion, and the claim is proved.

Note that the proof of Proposition 3.9 is at last complete. This also finally completes the proof of Theorem 3.3.

$\square$

We now discuss an interesting complement to Theorem 3.3, which is a variant of a result of Greenberg [G1], and which is very useful in the study of numerical examples. Our method of proof is different from that used by Greenberg; we state the most general result which emerges from our arguments.

3.11. Theorem. *Assume that*

(i) $p > 2$,

(ii) $\mathcal{S}(E/F)$ *is finite,*

(iii) *For each $P \neq 0$ in $E(F)(p)$, there exists a place v of F dividing p such that E has good reduction modulo v and the reduction of P modulo v is not zero.*

Then $H^1(\Gamma, \mathcal{S}(E/K_\infty)) = 0$.

3.12. Remarks. (i) Let A be any discrete p-primary Γ-module and let $X = \hat{A}$ be its Pontryagin dual. Assume that X is a finitely generated $\Lambda(\Gamma)$-module. Then $H^1(\Gamma, A) = 0$ is equivalent to the assertion that X has no finite non-zero Γ-submodules and that the characteristic power series (see Appendix A.1.5) of the $\Lambda(\Gamma)$-torsion submodule of X does not vanish at $T = 0$.

(ii) Note that condition (iii) of the theorem above is automatically true if $E(F)(p) = 0$.

(iii) Condition (iii) is also automatically valid if there exists a place v of F dividing p such that E has good reduction at v and the absolute ramification index e_v of v satisfies $e_v < p - 1$. For, it is well-known [Si, Chapter VII §3] that the formal group $\hat{E}_v$ of E at v will have a non-zero point of order p defined over F_v only if $e_v \geq p - 1$.

(iv) As we shall see in Chapter 5, some of the most interesting applications of Theorem 3.11 occur when $F = \mathbb{Q}(\mu_p)$, and so the ramification index of p in F is equal to $p - 1$.

(v) The results of this chapter can be used to give a simple proof that $\mathcal{S}(E/K_\infty) = 0$ under the following conditions. We assume that the hypotheses of Theorem 3.3 are valid for E/F and p. Further, we assume that the quantity $\rho_p(E/F)$ defined by (13) is equal to 1. Hence, by Theorem 3.3, we have $\chi(\Gamma, \mathcal{S}(E/K_\infty)) = 1$. Finally, we assume that condition (iii) of Theorem 3.11 is also valid for E/F and p. Thus Theorem 3.11 implies that $H^1(\Gamma, \mathcal{S}(E/K_\infty)) = 0$, whence $H^0(\Gamma, \mathcal{S}(E/K_\infty)) = 0$ because $\chi(\Gamma, \mathcal{S}(E/K_\infty)) = 1$. But, as $\mathcal{S}(E/K_\infty)$ is a discrete Γ-module, $H^0(\Gamma, \mathcal{S}(E/K_\infty)) = 0$ gives $\mathcal{S}(E/K_\infty) = 0$. This rather curious argument works very well for certain numerical examples, including $A_1 = X_1(11)$ and $p = 5$, as we shall see in Chapter 5.

3.13. Proof of Theorem 3.11. We consider the following commutative diagram with exact rows

$$
\begin{array}{ccc}
H^1(G_{S,\infty}, E_{p^\infty})^\Gamma & \xrightarrow{\phi_\infty} & \underset{v \in S}{\oplus}\, J_v(K_\infty)^\Gamma \\
\beta \uparrow & & \gamma \uparrow \\
H^1(G_S, E_{p^\infty}) & \xrightarrow{\ \ \lambda\ \ } \underset{v \in S}{\oplus}\, J_v(F) \xrightarrow{\ \theta\ } & \widehat{E(F)(p)} \longrightarrow 0.
\end{array}
\tag{39}
$$

Here the commutative square is simply the right hand part of the fundamental diagram (9), and the exactness of the bottom row follows from our hypotheses (i) and (ii) and Proposition 1.9. We claim that

$$\theta(\mathrm{Ker}\ \gamma) = \widehat{E(F)}(p). \tag{40}$$

Assuming (40), and using the fact that γ is surjective, a simple diagram chase shows that $\gamma \circ \lambda$ is then surjective. Hence the map ϕ_∞ must certainly be surjective because $\phi_\infty \circ \beta = \gamma \circ \lambda$. We now prove (40).

For each $v \in S$, let $E_U(K_{\infty,v})$ be the subgroup of universal norms in $E(F_v)$ as defined in **(2.11)**. Then we claim that (40) is equivalent to the assertion that the natural map

$$\tau\ :\ E(F)(p)\ \ \to\ \ \underset{v\in S}{\oplus}\, E(F_v)/E_U(K_{\infty,v})$$

is injective. This follows on noting that the dual of θ is, via the Tate pairing, the natural inclusion

$$E(F)(p)\ \ \hookrightarrow\ \ \underset{v\in S}{\oplus}\, E(F_v)^*,$$

and that the exact orthogonal complement of Ker γ under this pairing is $\underset{v\in S}{\oplus}\, E_U(K_{\infty,v})^*$ (note that $E(F_v)/E_U(K_{\infty,v})$ is automatically pro-p because it is dual to Ker γ_v). We now proceed to prove the injectivity of τ. Suppose, on the contrary, that there exists $P \neq 0$ in $E(F)(p)$ such that $\tau(P) = 0$. It follows that $P \in E_U(K_{\infty,v})$ for all places v of F dividing p. For each $n \geq 0$, let $K_{n,v}$ be the unique sub-extension of $K_{\infty,v}$ of degree p^n over F_v, and let $k_{n,v}$ be the residue field of $K_{n,v}$. Since v is assumed to divide p, it is a basic property of the cyclotomic $\mathbb{Z}_p$-extension K_∞ over F that there exists an integer n_0 such that $K_{\infty,v}/K_{n,v}$ is totally ramified for all $n \geq n_0$ (cf. Appendix). Hence $k_{n,v} = k_{n_0,v}$ for all $n \geq n_0$. As $P \in E_U(K_{\infty,v})$, we know that

$$P = N_{K_{n,v}/F_v}(P_n)$$

for some $P_n \in E(K_{n,v})$. We now use condition (iii) of the theorem, and choose v dividing p such that E has good reduction modulo v and such that the reduction $\tilde{P}$ of P modulo v is non-zero in $\tilde{E}_v(k_v)$. We have the commutative diagram

$$
\begin{array}{ccc}
E(K_{n,v}) & \longrightarrow & \tilde{E}_v(k_{n,v}) \\
{\scriptstyle N_{K_{n,v}/F_v}}\downarrow & & \downarrow{\scriptstyle \widetilde{N_{K_{n,v}/F_v}}} \\
E(F_v) & \longrightarrow & \tilde{E}_v(k_v),
\end{array}
\tag{41}
$$

where the left vertical map is the norm map, and the right vertical map is its reduction modulo v, and the horizontal maps are reduction modulo v. Now choose n so large such that p^{n-n_0} annihilates the p-primary subgroup of $\tilde{E}_v(k_{n,v})$. Since the norm map from $K_{n,v}$ to $K_{n_0,v}$ is multiplication by p^{n-n_0} on $\tilde{E}_v(k_{n,v})$, it follows that the reduction modulo v of P_n is sent to an element R of $\tilde{E}_v(k_v)$ of order prime to p by the right vertical map of (41). But, by the commutativity of (41), $R = \tilde{P}$. But $\tilde{P}$ is non-zero of p-power order, this is a contradiction, and the proof of Theorem 3.11 is complete.

$\square$

THE DIVISION FIELD CASE

3.14. In the rest of the chapter, we outline, without proofs, how far we can go towards computing the Euler characteristic of the Selmer group over the field $F_\infty = F(E_{p^\infty})$. We do not give proofs, and refer the interested reader to [C], [C-H] and [H] for these. The results described here are joint work of the first author with Susan Howson. For the remainder of this chapter, we shall assume that E does not admit complex multiplication. Indeed, the case when E admits complex multiplication is less interesting and essentially well-known. We again write Σ for the Galois group of F_∞ over F, so that Σ can be identified with a subgroup of $GL_2(\mathbb{Z}_p)$, which is open by Serre's theorem. We assume that $p \geq 5$, which guarantees that Σ has no non-trivial p-torsion. Hence, by [Se3], Σ has p-cohomological dimension equal to 4. If A is a discrete p-primary Σ-module, we shall say that A has finite Σ-Euler characteristic if $H^i(\Sigma, A)$ is finite for $i = 0, \ldots, 4$, and we then define its Euler characteristic by

$$\chi(\Sigma, A) = \prod_{i=0}^{4} (\# \, H^i(\Sigma, A))^{(-1)^i}. \tag{42}$$

Our aim is to compute $\chi(\Sigma, \mathcal{S}(E/F_\infty))$. Unfortunately, no unconditional analogue of Theorem 3.3 has been proved to date, and so we begin by stating a conjecture (see [C-H], [H]).

Let v be any place of F. As before, (cf. **(3.2)**), let

$$c_v = [E(F_v) : E_0(F_v)], \qquad d_v = \# \, \tilde{E}_v(k_v)$$

where k_v is the residue field of v. Let $B = B(E)$ be the (possibly empty) set of places v of F where $\mathrm{ord}_v(j_E) < 0$; here j_E denotes the j-invariant of E. For each $v \in B$, let $L_v(E, s)$ denote the Euler factor at v of the complex L-function of E over F. Hence $L_v(E, s)$ is 1, $(1/(1 - Nv)^s)$ or $(1/(1 + Nv)^s)$ according as E has additive reduction at v, split multiplicative reduction at v, or non-split multiplicative reduction at v. Assuming $\mathcal{S}(E/F)$ is finite, we define

$$\xi_p(E/F) = \frac{\#\, \text{III}(E/F)(p)}{\#\, (E(F)(p))^2} \times \prod_{v|p}(d_v^{(p)})^2 \times \prod_{v\in B}(c_v/L_v(E,1))^{(p)}. \tag{43}$$

Note that, since $p \geq 5$, $c_v^{(p)} = 1$ for all $v \notin B$ [Si], and so $\xi_p(E/F)$ is related to $\rho_p(E/F)$ (cf. (13)) by the formula

$$\xi_p(E/F) = \rho_p(E/F) \times \prod_{v\in B}(1/L_v(E,1))^{(p)}. \tag{44}$$

3.15. Conjecture. *Assume that*

(i) $p \geq 5$,

(ii) *E has good ordinary reduction at all places v of F dividing p,*

(iii) *$\mathcal{S}(E/F)$ is finite.*

Then $\mathcal{S}(E/F_\infty)$ has finite Σ-Euler characteristic given by

$$\chi(\Sigma, \mathcal{S}(E/F_\infty)) = \xi_p(E/F).$$

We now explain the partial results which we can prove in the direction of this conjecture. Let X be a finitely generated left $\Lambda(\Sigma)$-module. Recall that the $\Lambda(\Sigma)$-rank of X is defined in **2.15**. We say that X is $\Lambda(\Sigma)$-*torsion* if it has $\Lambda(\Sigma)$-rank equal to zero. Equivalently, if we choose an open subgroup Φ of Σ such that $\Lambda(\Phi)$ is Noetherian and has no zero-divisors, then X is $\Lambda(\Sigma)$-torsion if and only if every element of X has a non-zero annihilator in $\Lambda(\Phi)$.

3.16. Theorem. *[C-H] Assume the hypotheses of Conjecture 3.15. If $\widehat{\mathcal{S}(E/F_\infty)}$ is $\Lambda(\Sigma)$-torsion, then $\mathcal{S}(E/F_\infty)$ has finite Euler characteristic given by*

$$\chi(\Sigma, \mathcal{S}(E/F_\infty)) = \xi_p(E/F).$$

Moreover, $H^i(\Sigma, \mathcal{S}(E/F_\infty)) = 0$ for $i = 2, 3, 4$.

We recall the exact sequence, where now $G_{S,\infty} = G(F_S/F_\infty)$,

$$0 \longrightarrow \mathcal{S}(E/F_\infty) \longrightarrow H^1(G_{S,\infty}, E_{p^\infty}) \xrightarrow{\;\lambda_\infty\;} \underset{v \in S}{\oplus} J_v(F_\infty).$$

Assume that E has good ordinary reduction at all primes v of F dividing p. Then it can be shown that $\widehat{\mathcal{S}(E/F_\infty)}$ is $\Lambda(\Sigma)$-torsion if and only if λ_∞ is surjective (cf. [C]). However, we hasten to point out that, at present, we only know how to prove $\widehat{\mathcal{S}(E/F_\infty)}$ is $\Lambda(\Sigma)$-torsion for a few isogeny classes of elliptic curves E over $\mathbb{Q}$ with $p = 5$ or $p = 7$, including the three elliptic curves A_0, A_1, A_2 of conductor 11, with $p = 5$ (cf. **2.18**), even though we conjecture this to be true whenever $r_p(E/F) = 0$. We remark that an earlier claim by Harris [H1] to prove such examples is unjustified (see [B-H]), because the analogue of Lemma 2.10 is false when Γ is replaced by an open subgroup of $GL_2(\mathbb{Z}_p)$. As for unconditional partial results in the direction of Conjecture 3.15, the best we can do is the following. Let

$$\phi_\infty \;:\; H^1(G_{S,\infty}, E_{p^\infty})^\Sigma \longrightarrow \underset{v \in S}{\oplus} J_v(F_\infty)^\Sigma \tag{45}$$

be the map which is induced by the localization maps λ_∞. We remark that by a result of Serre [Se4], the groups $H^i(\Sigma, E_{p^\infty})$ are all finite.

3.17. Theorem. *[C-H] Assume that*

(i) *$p \geq 5$,*

(ii) *E has good ordinary reduction at all places v of F dividing p,*

(iii) *$\mathcal{S}(E/F)$ is finite.*

Then Coker ϕ_∞, *$H^0(\Sigma, \mathcal{S}(E/F_\infty))$ and $H^1(\Sigma, \mathcal{S}(E/F_\infty))$ are all finite. Moreover, we have*

(a) *$\# \, H^0(\Sigma, \mathcal{S}(E/F_\infty)) = \xi_p(E/F) \times \# \, H^3(\Sigma, E_{p^\infty}) \times \#$ Coker ϕ_∞,*

(b) *$\# \, H^1(\Sigma, \mathcal{S}(E/F_\infty))$ divides $\# \, H^3(\Sigma, E_{p^\infty}) \times \#$ Coker ϕ_∞.*

At present, we do not know how to prove even the finiteness of $H^i(\Sigma, \mathcal{S}(E/F_\infty))$ ($i = 2, 3, 4$) under the hypotheses of Theorem 3.17 (of course, the finiteness of these cohomology groups is known when $\widehat{\mathcal{S}(E/F_\infty)}$ is $\Lambda(\Sigma)$-torsion, and as mentioned above, there are numerical examples where this holds). We also do not know any examples where we can prove that Coker ϕ_∞ is not zero, although we suspect that such examples probably exist. However, when $F = \mathbb{Q}$, we can prove the following stronger result.

3.18. Theorem. *[C-H] Let $E/\mathbb{Q}$ be a modular elliptic curve whose complex L-function $L(E, s)$ does not vanish at $s = 1$. Let p be a prime ≥ 5, where E has good ordinary reduction. Then* Coker $\phi_\infty = 0$, *where ϕ_∞ is given by (45). Moreover, we have $H^0(\Sigma, \mathcal{S}(E/F_\infty))$ and $H^1(\Sigma, \mathcal{S}(E/F_\infty))$ are both finite, and*

(a) $\# H^0(\Sigma, \mathcal{S}(E/F_\infty)) = \xi_p(E/F) \times \# H^3(\Sigma, E_{p^\infty})$,

(b) $\# H^1(\Sigma, \mathcal{S}(E/F_\infty))$ *divides* $\# H^3(\Sigma, E_{p^\infty})$.

We note that Kolyvagin's deep theorem tells us that, for modular E over $\mathbb{Q}$, $L(E, 1) \neq 0$ implies that $\mathcal{S}(E/\mathbb{Q})$ is finite for all primes p.

Finally we mention the following result, which is used in the proof of Theorems 3.16, 3.17 and 3.18, which we shall prove in the Appendix (see **A.2.9**). Let E/F be any elliptic curve and p be a prime such that the Galois group Σ has no element of order p (this is automatically true if $p \geq 5$). Let d be the dimension of Σ as a p-adic Lie group. Thus $d = 4$ if E has no complex multiplication and by the theory of complex multiplication, $d = 2$ when E admits complex multiplication. We define

$$\chi(\Sigma, E_{p^\infty}) = \prod_{i=0}^{d} (\# H^i(\Sigma, E_{p^\infty}))^{(-1)^i}.$$

3.19. Theorem. *We have $\chi(\Sigma, E_{p^\infty}) = 1$ and $H^i(\Sigma, E_{p^\infty}) = 0$ for all $i \geq d$.*

Note that the case $p = 2$ is allowed in Theorem 3.19. This result was originally proven by Serre [Se5] when E/F has no complex multiplication and $p > 5$. In the Appendix, we give a different and simpler proof to Serre, which easily generalizes to arbitrary abelian varieties defined over a number field (see [C-S]).

Chapter 4

Numerical Examples over the Cyclotomic $\mathbb{Z}_p$-extension of $\mathbb{Q}$

INTRODUCTION

4.1. In this chapter, we discuss some numerical examples of elliptic curves over $\mathbb{Q}$, which illustrate the general theory developed in Chapters 2 and 3. We study the Iwasawa theory of these curves over the cyclotomic $\mathbb{Z}_p$-extension $\mathbb{Q}_\infty$ of $\mathbb{Q}$. As usual, we write $\Gamma = G(\mathbb{Q}_\infty/\mathbb{Q})$, and let $\Lambda(\Gamma)$ denote the Iwasawa algebra of Γ. We work with the single isogeny class of the curves of conductor 11, and one isogeny class of curves of conductor 294. We are grateful to T. Fisher [F] for pointing out to us the latter isogeny class, and providing us with data about a 7-descent on it.

The isogeny class of curves of conductor 11 was already introduced in Chapter 2. We recall that these three curves are

$$
\begin{aligned}
A_0 : y^2 + y &= x^3 - x^2 - 10x - 20 \\
A_1 : y^2 + y &= x^3 - x^2 \\
A_2 : y^2 + y &= x^3 - x^2 - 7820x - 263580.
\end{aligned}
$$

The curve A_0 is the modular curve $X_0(11)$, and A_1 is the modular curve $X_1(11)$. All three curves are isogenous over $\mathbb{Q}$, the isogenies being of degree 5. When there is no need to distinguish between the three curves, we shall simply write A to denote any one of them. The complex L-function of A over $\mathbb{Q}$ is given by

$$
L(A, s) = \sum_{n=1}^{\infty} a_n/n^s,
$$

"

where the integers a_n $(n = 1, 2, \dots)$ are defined by the expansion

$$q \prod_{n=1}^{\infty} (1 - q^n)^2 \prod_{n=1}^{\infty} (1 - q^{11n})^2 = \sum_{n=1}^{\infty} a_n q^n.$$

This is the q-series expansion of a cusp form of weight two for the subgroup $\Gamma_0(11)$ of $SL_2(\mathbb{Z})$ consisting of all matrices

$$\begin{pmatrix} a & b \\ c & d \end{pmatrix}$$

with a, b, c, d in $\mathbb{Z}$, $ad - bc = 1$ and $c \equiv 0 \mod 11$. It is easily verified by computation (cf. [Cr]) that $L(A, 1) \neq 0$. Hence, by Kolyvagin's theorem, it follows that $A_i(\mathbb{Q})$ and $\text{III}(A_i/\mathbb{Q})$ are both finite for $0 \leq i \leq 2$.

4.2. The three curves A_0, A_1 and A_2 can be distinguished by the structure of their groups of 5-division points considered as Galois modules for the Galois group $G = G(\overline{\mathbb{Q}}/\mathbb{Q})$.

On A_0, the point $(5, 5)$ is a non-trivial 5-division point and we have

$$A_{0,5} = \mathbb{Z}/5\mathbb{Z} \oplus \mu_5 \tag{46}$$

as G-modules. On A_1, the point $(0, 0)$ is a non-trivial 5-division point, and we have the exact sequence of G-modules

$$0 \to \mathbb{Z}/5\mathbb{Z} \to A_{1,5} \to \mu_5 \to 0 \tag{47}$$

the quotient being μ_5 by virtue of the Weil pairing. This exact sequence is non-split as a sequence of $G(\overline{\mathbb{Q}}_{11}/\mathbb{Q}_{11})$-modules because the discriminant $\Delta(A_1)$ is -11. Indeed, A_1 has split multiplicative reduction at 11, and hence A_1 is a Tate curve over $\mathbb{Q}_{11}$ (cf. Appendix A.3). We have

$$A_1(\overline{\mathbb{Q}}_{11}) = \overline{\mathbb{Q}}_{11}^* / q_{A_1}^{\mathbb{Z}},$$

where q_{A_1} is the Tate period of of A_1 at 11 (cf. Appendix A.3). If the whole group of 5-division points were rational over a finite extension L of $\mathbb{Q}_{11}$, then the absolute ramification index of L would have to be divisible by 5 because q_{A_1} has order 1 at 11. In particular, this proves that (47) cannot split over $\mathbb{Q}_{11}(\mu_5) = \mathbb{Q}_{11}$. On A_2, there are no non-trivial rational points, and we have the exact sequence of G-modules

$$0 \longrightarrow \mu_5 \longrightarrow A_{2,5} \longrightarrow \mathbb{Z}/5\mathbb{Z} \longrightarrow 0. \tag{48}$$

Again this sequence is not split as a sequence of $G(\overline{\mathbb{Q}}_{11}/\mathbb{Q}_{11})$-modules as the discriminant $\Delta(A_2) = -11$. We note that $A_1 = A_0/\mu_5$ and $A_2 = A_0/(\mathbb{Z}/5\mathbb{Z})$. Also, $A_0 = A_2/\mu_5$, and so we see that $A_{2,5^\infty}$ contains a cyclic subgroup of order 25 which is invariant under G.

4.3. The second isogeny class of elliptic curves is given by the following two curves of conductor 294:

$$
\begin{aligned}
B_1 : y^2 + xy &= x^3 - x - 1 \\
B_2 : y^2 + xy &= x^3 - 141x + 657.
\end{aligned}
$$

These are the curves $B1$ and $B2$ of conductor 294 in Cremona's tables [Cr]. In making his tables, Cremona has verified that all isogeny classes listed are complete, and so there are no other elliptic curves over $\mathbb{Q}$ which are isogenous to B_1 and B_2. The discriminants of the curves are given by

$$
\Delta(B_1) = -2 \cdot 3 \cdot 7^2, \ \Delta(B_2) = -2^7 \cdot 3^7 \cdot 7^2.
$$

Cremona's tables show that neither B_1 nor B_2 possesses any rational point of order prime to 7. This can also be seen from the fact that the number of points modulo 5 or modulo 11 is 7. From Cremona [Cr], we see that B_2 has a rational point of order 7, and that $B_1 = B_2/(\mathbb{Z}/7\mathbb{Z})$, where $\mathbb{Z}/7\mathbb{Z}$ denotes the subgroup generated by the point of order 7. Both curves have split multiplicative reduction at 2 and 3, and bad reduction at 7. In fact, both curves achieve good ordinary reduction over the field $\mathbb{Q}_7(\mu_7)$. This can be seen explicitly for B_1 as follows (and then it follows automatically for the isogenous curve B_2). If we make the change of variables

$$
x = x' + 4, \ y = y' + 3x' + 5,
$$

we obtain the new equation for B_1

$$
y^2 + 7xy + 14y = x^3 + 14. \tag{49}
$$

Making the change of variables, $x = \pi^2 x'$, $y = \pi^3 y'$, where π is any local parameter of $\mathbb{Q}_7(\mu_7)$, we clearly obtain an equation for B_1 over $\mathbb{Q}_7(\mu_7)$ which has good reduction because 7 has ramification index 6 in $\mathbb{Q}_7(\mu_7)$. One also verifies easily that, on taking $\pi = 1 - \zeta$, where ζ is a primitive 7-th root of unity, the reduction of this last equation is the curve

$$
y^2 = x^3 - 2 \tag{50}
$$

over the field $\mathbb{F}_7$. The number of points on this last curve over $\mathbb{F}_7$ is equal to 7, and so it is ordinary. Let Φ denote the Galois group of the totally ramified extension $\mathbb{Q}_7(\mu_7)$ over $\mathbb{Q}_7$. Now the theory of the Néron model shows that Φ injects into the automorphism group of (50) over $\mathbb{F}_7$ (see [Se2, §5.6]), and so it must coincide with the whole automorphism group, since the latter has order 6. But the automorphism group of (50) is generated by $(x, y) \mapsto (\omega x, -y)$ where ω is a primitive cube root of unity, and so it certainly does not act trivially on the $\mathbb{F}_7$-rational points of (50). We conclude that the Galois group Φ also acts non-trivially on the $\mathbb{F}_7$-rational points of (50), a fact which we shall use later. We thank B. Totaro for making this very useful remark.

The structure of the groups of 7-division points on B_1 and B_2 are given by the exact sequences of G-modules

$$0 \longrightarrow \mu_7 \longrightarrow B_{1,7} \longrightarrow \mathbb{Z}/7\mathbb{Z} \longrightarrow 0, \tag{51}$$

$$0 \longrightarrow \mathbb{Z}/7\mathbb{Z} \longrightarrow B_{2,7} \longrightarrow \mu_7 \longrightarrow 0. \tag{52}$$

Since B_1 and B_2 form a complete isogeny class over $\mathbb{Q}$, and neither B_1 nor B_2 admit complex multiplication, we see that neither of these exact sequences can split as G-modules. Of course, one could also use a local argument at the primes 2 or 3 to prove this for B_1, but not for B_2 because $\Delta(B_2) = -2^7 \cdot 3^7 \cdot 7^2$.

Finally, we note that B_1 and B_2 are both modular, and their complex L-function does not vanish at $s = 1$ (see [Cr]). Hence Kolyvagin's theorem tells us that $B_i(\mathbb{Q})$ and $\text{III}(B_i(\mathbb{Q}))$ are finite for $i = 1, 2$.

IWASAWA THEORY OVER $\mathbb{Q}_\infty$ FOR THE CURVES OF CONDUCTOR 11

4.4. The study of the Iwasawa theory of the curves of conductor 11 was begun by Mazur in his seminal paper [Ma]. The results we describe now are essentially folklore (see [G2]) and have been well-known to the experts for quite some time.

In describing the Iwasawa theory for a given prime p, we shall assume that

$$\text{III}(A_i/\mathbb{Q})(p) = 0 \ (0 \le i \le 2). \tag{53}$$

This is in accord with the conjecture of Birch and Swinnerton-Dyer, which predicts that the Tate-Shafarevich group of each A_i over $\mathbb{Q}$ is trivial (cf. [Cr]). It should be possible to prove (53) for all primes p by computing non-trivial Heegner points and using the formula of Gross-Zagier (cf. [Gr]), but the details do not appear in the literature. We let $\Gamma = G(\mathbb{Q}_\infty/\mathbb{Q})$ and recall that $\Lambda(\Gamma)$ denotes the Iwasawa algebra of Γ. We shall also let A denote any one of the three curves A_i, $0 \le i \le 2$, and take the set S to be $\{p, 11\}$.

4.5. Theorem. *Let p be an odd prime such that (53) holds. Assume that A has good supersingular reduction at p (e.g., $p = 19, 29, \dots$). Then*

$$\widehat{\mathcal{S}(A/\mathbb{Q}_\infty)} \simeq \Lambda(\Gamma)$$

as a $\Lambda(\Gamma)$-module.

Proof. Let $X = \widehat{\mathcal{S}(A/\mathbb{Q}_\infty)}$. By Theorem 2.14, we know that

$$\Lambda(\Gamma) \text{ - rank of } X = 1.$$

We claim that

$$\mathcal{S}(A/\mathbb{Q}_\infty)^\Gamma \simeq \mathbb{Q}_p/\mathbb{Z}_p. \tag{54}$$

In order to prove the claim, we simply analyse the fundamental diagram (9) with $F = \mathbb{Q}$ and $H_\infty = \mathbb{Q}_\infty$:

$$
\begin{array}{ccccccc}
0 & \longrightarrow & \mathcal{S}(A/\mathbb{Q}_\infty)^\Gamma & \longrightarrow & H^1(G_{S,\infty}, A_{p^\infty})^\Gamma & \xrightarrow{\phi_\infty} & \underset{v \in S}{\oplus} J_v(\mathbb{Q}_\infty)^\Gamma \\
 & & \big\uparrow{\scriptstyle \alpha} & & \big\uparrow{\scriptstyle \beta} & & \big\uparrow{\scriptstyle \gamma} \\
0 & \longrightarrow & \mathcal{S}(A/\mathbb{Q}) & \longrightarrow & H^1(G_S, A_{p^\infty}) & \xrightarrow{\lambda} & \underset{v \in S}{\oplus} J_v(\mathbb{Q}).
\end{array}
$$

Observe that $A(\mathbb{Q})(p)$ must be zero because this group injects into $\tilde{A}(\mathbb{F}_5) = \mathbb{Z}/5\mathbb{Z}$ under reduction modulo 5. By virtue of (53), and the fact that $A(\mathbb{Q})(p) = 0$, we have $\mathcal{S}(A/\mathbb{Q}) = 0$. Thus it follows that the map λ above is surjective as Coker $\lambda \simeq \widehat{A(\mathbb{Q})(p)}$ by Proposition 1.9. Note also that β is injective because Ker $\beta = H^1(\Gamma, A_{p^\infty}(\mathbb{Q}_\infty))$ and $A_{p^\infty}(\mathbb{Q}_\infty) = 0$ since $H^0(\Gamma, A_{p^\infty}(\mathbb{Q}_\infty)) = 0$. As the map β is always surjective, we deduce that it is an isomorphism in this case. In view of these remarks, we conclude that

$$\text{Ker } \gamma \simeq \mathcal{S}(A/\mathbb{Q}_\infty)^\Gamma. \tag{55}$$

As $S = \{p, 11\}$, and we have $\gamma = \gamma_p \oplus \gamma_{11}$. Now

$$c_{11}(A_i) = \begin{cases} 5 & \text{for} \ \ i = 0 \\ 1 & \text{for} \ \ 1 \leq i \leq 2. \end{cases}$$

Hence it follows from Lemma 3.4 that γ_{11} is injective. We are thus left with analysing γ_p.

We prove that γ_p is in fact the zero map. Indeed, if w denotes the unique prime of $\mathbb{Q}$ above p, we have

$$H^1(\mathbb{Q}_{\infty,w}, A)(p) = 0,$$

by Proposition 4.8 of [C-G]. Note that since A has supersingular reduction at p, the Galois submodule C of A_{p^∞} defined in [C-G, p. 150] is equal to the whole of A_{p^∞}. Thus

$$\mathrm{Ker} \ \gamma_p = H^1(\mathbb{Q}_p, A)(p)$$

which is dual to $A(\mathbb{Q}_p)^*$ by Tate local duality. Our claim (54) now follows on observing that $A(\mathbb{Q}_p)^*$ is a free $\mathbb{Z}_p$-module of rank 1. Indeed, any torsion element of $A(\mathbb{Q}_p)^*$ will have to lie on the formal group $\widehat{A}$, but $\widehat{A}$ is torsion free as the ramification index of $\mathbb{Q}_p$ is $< p - 1$ (cf. [Si, Chapter VII]). The final part of the proof consists of a simple argument with $\Lambda(\Gamma)$-modules, (cf. [Appendix A.1]) along with (54) and the fact that X has $\Lambda(\Gamma)$-rank equal to 1. Let Y denote the $\Lambda(\Gamma)$-torsion submodule of X, so that we have the exact sequence

$$0 \longrightarrow Y \longrightarrow X \longrightarrow X/Y \longrightarrow 0$$

of $\Lambda(\Gamma)$-modules. Now $(X/Y)^\Gamma = 0$ because X/Y has no non-zero $\Lambda(\Gamma)$-torsion, and thus we obtain the exact sequence

$$0 \longrightarrow Y_\Gamma \longrightarrow X_\Gamma \longrightarrow (X/Y)_\Gamma \longrightarrow 0. \qquad (56)$$

But $X_\Gamma = \mathbb{Z}_p$ because of (54) and the fact that X_Γ is dual to $\mathcal{S}(A/\mathbb{Q}_\infty)^\Gamma$. Also X/Y has $\Lambda(\Gamma)$-rank equal to 1, and thus

$$\mathbb{Z}_p - \text{rank of } (X/Y)_\Gamma \geq 1.$$

We conclude from (55) that Y_Γ is finite. Hence $Y_\Gamma = 0$ because it injects into $X_\Gamma \simeq \mathbb{Z}_p$. But Y is a compact $\Lambda(\Gamma)$-module, and so

$$Y_\Gamma = 0 \Longrightarrow Y = 0.$$

The structure theory for finitely generated $\Lambda(\Gamma)$-modules gives (cf. Appendix A.1) an exact sequence

$$0 \longrightarrow X \longrightarrow \Lambda(\Gamma) \longrightarrow B \longrightarrow 0,$$

where B is a finite $\Lambda(\Gamma)$-module. Taking Γ-invariants of this last exact sequence, we obtain the exact sequence

$$0 \longrightarrow B^\Gamma \longrightarrow X_\Gamma \longrightarrow \mathbb{Z}_p \longrightarrow B_\Gamma \longrightarrow 0.$$

But X_Γ is a free $\mathbb{Z}_p$-module of rank 1, and so $B^\Gamma = 0$ because it is finite. This proves in turn that $B = 0$, and the proof of the theorem is now complete.

$\square$

4.6. Theorem. *Let $p \neq 5$ be an odd prime satisfying (53). Assume that A has good ordinary reduction at p (e.g., $p = 3, 7, 13, \ldots$). Then $\mathcal{S}(A/\mathbb{Q}_\infty) = 0$.*

Proof. The proof involves analysing the fundamental diagram as before. Since A has good ordinary reduction at p, by Theorem 3.3, the Euler characteristic $\chi(\Gamma, \mathcal{S}(A/\mathbb{Q}_\infty))$ is finite, and

$$\chi(\Gamma, \mathcal{S}(A/\mathbb{Q}_\infty)) = \rho_p(A/\mathbb{Q})$$

where $\rho_p(A/\mathbb{Q})$ is defined by (13). We now proceed to calculate $\rho_p(A/\mathbb{Q})$. Let $\tilde{A}_p$ denote the reduced curve over $\mathbb{F}_p$. Then we have

$$\#\ \tilde{A}_5(\mathbb{F}_5) = 5, \ \ \tilde{A}_p(\mathbb{F}_p)(p) = 0 \ for \ p \neq 5.$$

This latter assertion follows from Hasse's bound on the order of $\tilde{A}_p(\mathbb{F}_p)$ and the fact that 5 must divide the order of $\tilde{A}_p(\mathbb{F}_p)$ for all $p \neq 11$. We also have

$$c_{11}(A_0) = 5, \ c_{11}(A_i) = 1 \ for \ 1 \leq i \leq 2.$$

Since $p \neq 5$, we again have $A(\mathbb{Q})(p) = 0$. Hence assuming (53), we deduce from (13) that $\rho_p(A/\mathbb{Q}) = 1$. This implies that

$$\chi(\Gamma, \mathcal{S}(A/\mathbb{Q}_\infty)) = 1.$$

On the other hand, since $A(\mathbb{Q})(p) = 0$, Theorem 3.11 shows that

$$H^1(\Gamma, \mathcal{S}(A/\mathbb{Q}_\infty)) = 0.$$

This in turn implies that $H^0(\Gamma, \mathcal{S}(A/\mathbb{Q}_\infty)) = 0$ and hence $\mathcal{S}(A/\mathbb{Q}_\infty) = 0$, as $\mathcal{S}(A/\mathbb{Q}_\infty)$ is a discrete Γ-module. The theorem is therefore proved.

$\square$

4.7. Theorem. *For $p = 5$, we have*

$$
\begin{aligned}
\mathcal{S}(A_1/\mathbb{Q}_\infty) &= 0 \\
\mathcal{S}(\widehat{A_0/\mathbb{Q}_\infty}) &= \Lambda(\Gamma)/(5) \\
\mathcal{S}(\widehat{A_2/\mathbb{Q}_\infty}) &= \Lambda(\Gamma)/(5^2).
\end{aligned}
$$

Proof. We will only sketch the proof of this theorem, as it is well-known, and treated in detail in [G2]. In the case $p = 5$, we have

$$
\#\, A_0(\mathbb{Q})(5) = 5, \ \#\, A_1(\mathbb{Q})(5) = 5, \ \#\, A_2(\mathbb{Q}) = 0.
$$

Further, it can be shown by a 5-descent argument [G2] that

$$
\text{III}(A_i(\mathbb{Q}))(5) = 0 \ for \ all \ i, \ 0 \le i \le 2.
$$

Hence we deduce from Theorem 3.3 that

$$
\chi(\Gamma, \mathcal{S}(A_i/\mathbb{Q}_\infty)) = \begin{cases} 5 & \text{for } i = 0 \\ 1 & \text{for } i = 1 \\ 5^2 & \text{for } i = 2. \end{cases}
$$

Moreover, since there is no 5-torsion on the formal group of A_i over $\mathbb{Q}$, Theorem 3.11 shows that

$$
H^1(\Gamma, \mathcal{S}(A_i/\mathbb{Q}_\infty)) = 0 \ for \ all \ i, \ 0 \le i \le 2.
$$

Therefore it follows that

$$
H^0(\Gamma, \mathcal{S}(A_1/\mathbb{Q}_\infty)) = 0
$$

and so $\mathcal{S}(A_1/\mathbb{Q}_\infty) = 0$, as it is a discrete p-primary Γ-module.

Let $f_i(T)$ denote a characteristic power series for $\mathcal{S}(\widehat{A_i/\mathbb{Q}_\infty})$ as a $\Lambda(\Gamma)$-module (see Appendix); here we are identifying $\Lambda(\Gamma)$ with the ring $\mathbb{Z}_p[[T]]$ of formal power series in T with coefficients in $\mathbb{Z}_p$. Hence we conclude from Propostion A.1.7 of the Appendix that

$$
f_0(0) = 5u_0, \ f_2(0) = 5^2 u_2 \tag{57}
$$

where u_0 and u_2 are 5-adic units. There are now two ways to complete the proof of Theorem 4.7. The first is to invoke Proposition 5.7 of [G2], which is applicable because A_0 and A_2 have good ordinary reduction at 5, and $A_{i,5\infty}$ contains a non-trivial subgroup D_i which is stable under the action of G. In the case of A_0, $D_0 = \mu_5$, and so we deduce from Proposition 5.7 of [G2] that $f_0(T)$ is divisible by 5, whence it is clear from the first equation in (57), that we can take $f_0(T) = 5$. In the case of A_2, D_2 has order 25, and the action of complex conjugation on D_2 is odd because $\mu_5 \subset D_2$. We conclude in this second case that $f_2(T)$ is divisible by 5^2, and so by (57), we can take $f_2(T) = 5^2$. Alternatively, we could use the results of Perrin-Riou [P-R2] or Schneider [Sch], giving the change in the characteristic power series of the dual of Selmer under an isogeny. Since, we know that the characteristic power series of $\mathcal{S}(\widehat{A_1/\mathbb{Q}_\infty})$ is 1, we deduce also in this way that we can take $f_0(T) = 5$, $f_2(T) = 5^2$.

As $H^1(\Gamma, \mathcal{S}(A_0/\mathbb{Q}_\infty)) = 0$, it follows that $\mathcal{S}(\widehat{A_0/\mathbb{Q}_\infty})$ has no non-zero finite $\Lambda(\Gamma)$-submodule. Hence, since $f_0(T) = 5$, the structure theory for $\Lambda(\Gamma)$-modules implies that there is an embedding of $\mathcal{S}(\widehat{A_0/\mathbb{Q}_\infty})$ in $\Lambda(\Gamma)/5$ with finite cokernel. But $\Lambda(\Gamma)/5$ is the ring $\mathbb{F}_5[[T]]$, and so is a discrete valuation ring. Thus the image of $\mathcal{S}(\widehat{A_0/\mathbb{Q}_\infty})$ in $\Lambda(\Gamma)/5$ must be a principal ideal and so itself isomorphic to $\Lambda(\Gamma)/5$. This proves the assertion of Theorem 4.7 for A_0. We omit the detailed proof of Theorem 4.7 for A_2 and refer the reader to [G2] for a full explanation.

$\square$

4.8. To avoid possible confusion with our notation for the p-adic numbers, we write $\mathbb{Q}^{(n)}$ for the unique subfield of $\mathbb{Q}_\infty$ of degree p^n over $\mathbb{Q}$, and put $\Gamma_n = G(\mathbb{Q}_\infty/\mathbb{Q}^{(n)})$. We are still totally ignorant about the behaviour of the p-primary subgroup of $\text{III}(A)$ in the tower $\mathbb{Q}_\infty$ when A has good supersingular reduction at p. It does not appear to be known whether $\text{III}(A/\mathbb{Q}^{(n)})(p)$ $(n = 1, 2, \dots)$ are even finite. Moreover, assuming the finiteness, nothing seems to be known about the asymptotic behaviour of the order of $\text{III}(A/\mathbb{Q}^{(n)})(p)$ as $n \to \infty$. By contrast, it is interesting to note that one can obtain information about the arithmetic of the elliptic curves A_i over the finite layers of $\mathbb{Q}_\infty/\mathbb{Q}$ from Theorem 4.5 when p is an ordinary prime, but we restrict our discussion for simplicity, to $p \neq 5$. Again A denotes any of the three curves A_i.

4.9. Theorem. *Let p be any ordinary prime $\neq 5$ such that (53) is valid. Then $A(\mathbb{Q}^{(n)})$ is finite of order prime to p and $\text{III}(A/\mathbb{Q}^{(n)})(p) = 0$ for all $n \geq 0$.*

Proof. By Theorem 4.6, we have $\mathcal{S}(A/\mathbb{Q}_\infty) = 0$. Applying the fundamental diagram (9) with base field $F = \mathbb{Q}^{(n)}$, and noting that we have

$A(\mathbb{Q}^{(n)})(p) = 0$ because $A(\mathbb{Q})(p) = 0$, we conclude that the map

$$\alpha_n \ : \ \mathcal{S}(A/\mathbb{Q}^{(n)}) \ \rightarrow \ \mathcal{S}(A/\mathbb{Q}_\infty)^{\Gamma_n} = 0$$

must be injective, whence $\mathcal{S}(A/\mathbb{Q}^{(n)}) = 0$. The assertions of the theorem follow from this.

$\square$

IWASAWA THEORY OVER $\mathbb{Q}_\infty$ FOR THE CURVES OF CONDUCTOR 294

4.10. In describing the Iwasawa theory of B_1 and B_2 over $\mathbb{Q}_\infty$ for a given prime p, we shall assume that

$$\text{III}(B_i/\mathbb{Q})(p) = 0 \ (i = 1, \ 2). \tag{58}$$

In fact, the conjecture of Birch and Swinnerton-Dyer predicts that B_1 and B_2 both have trivial Tate-Shafarevich group over $\mathbb{Q}$ (see [Cr]), and again this should be provable for all p using Heegner points (see [Gr]). Exactly analogous statements to Theorems 4.5, 4.6 and 4.9 hold for all primes $p \neq 2, 3, 7$ for both the curves B_1 and B_2, assuming (58) holds. As the proofs are entirely similar, we do not repeat them. However we now discuss the prime $p = 7$ for B_1 and B_2, where subtler arguments are required. In fact, (58) is known to be true for $p = 7$, thanks to calculations of T. Fisher [F]. We also remark that despite the fact that $B_{1,7}$ has a subgroup isomorphic to μ_7 as a G-module, Greenberg's Proposition 5.7 in [G2] does not apply to B_1 because B_1 has had additive reduction at 7 (if it did apply, we would have a positive μ-invariant for $\mathcal{S}(B_1/\mathbb{Q}_\infty)$). In fact, the following result is true.

4.11. Theorem. *We have* $\mathcal{S}(B_1/\mathbb{Q}_\infty) = 0$.

Proof. We can take $S = \{2, 3, 7\}$. As always, we use the fundamental diagram (9) with $F = \mathbb{Q}$ and $H_\infty = \mathbb{Q}_\infty$.

$$
\begin{array}{ccccccc}
0 & \longrightarrow & \mathcal{S}(B_1/\mathbb{Q}_\infty)^{\Gamma} & \longrightarrow & H^1(G_{S,\infty}, B_{1,7^\infty})^{\Gamma} & \overset{\phi_\infty}{\longrightarrow} & \underset{v \in S}{\oplus} J_v(\mathbb{Q}_\infty)^{\Gamma} \\
 & & \big\uparrow{\alpha} & & \big\uparrow{\beta} & & \big\uparrow{\gamma} \\
0 & \longrightarrow & \mathcal{S}(B_1/\mathbb{Q}) & \longrightarrow & H^1(G_S, B_{1,7^\infty}) & \overset{\lambda}{\longrightarrow} & \underset{v \in S}{\oplus} J_v(\mathbb{Q}).
\end{array}
$$

Note that λ is surjective, because $B_1(\mathbb{Q})$ has no 7-torsion. Also β is injective for the same reason, and is surjective as always. In addition, $\mathcal{S}(B_1/\mathbb{Q}) = 0$, by Fisher's calculations. Hence we deduce that

$$\mathcal{S}(B_1/\mathbb{Q}_\infty)^\Gamma \simeq \mathrm{Ker}\ \gamma.$$

We now show that γ is injective, and so it will follow that $\mathcal{S}(B_1/\mathbb{Q}_\infty) = 0$ as required. But γ_2 and γ_3 are injective because $c_2 = c_3 = 1$ for our curve B_1. Hence it remains to show that γ_7 is also injective. To do this, we appeal once again to the theory of [C-G], where the canonical exact sequence is given by

$$0 \longrightarrow C \longrightarrow B_{1,7\infty} \longrightarrow D \longrightarrow 0 \tag{59}$$

where D is the group of all 7-power division points on the curve (50) over $\mathbb{F}_7$. Here D is isomorphic to $\mathbb{Q}_7/\mathbb{Z}_7$ as an abelian group, but the inertial subgroup of the Galois group $G(\overline{\mathbb{Q}}_7/\mathbb{Q}_7)$ acts on D via the finite quotient $\Delta = G(\mathbb{Q}_7(\mu_7)/\mathbb{Q}_7)$, as described earlier in **4.3**. Note also that $B_1(\mathbb{Q}_7)$ is a pro-7 group because we have the exact sequence

$$0 \longrightarrow \hat{B}_1(\mathbb{Q}_7) \longrightarrow B_1(\mathbb{Q}_7) \longrightarrow \mathbb{F}_7 \longrightarrow 0,$$

where $\hat{B}_1$ denotes the formal group of B_1 at 7 (this is because $c_7 = 1$ and the Kodaira symbol of the reduction at 7 is II). We shall show that every element of $B_1(\mathbb{Q}_7)$ is a universal norm for $\mathbb{Q}_{\infty,7}$, by using Theorem 5.2 of [C-G]. Observe that the map $\pi_{\mathbb{Q}_7}$ of [C-G] is surjective because B_1 has potential good reduction at 7. Thus Theorem 5.2 of [C-G] shows that the universal norm subgroup of $B_1(\mathbb{Q}_7)$ is dual to $H^1(\mathbb{Q}_7, D)/\mathrm{Ker}\ \rho$, where

$$\rho : H^1(\mathbb{Q}_7, D) \to H^1(\mathbb{Q}_{\infty,7}, D)$$

is the restriction map. Put

$$L = \mathbb{Q}_7(\mu_7),\, L_\infty = \mathbb{Q}_7(\mu_{7\infty}).$$

Now ρ is injective because its kernel is contained in $D(L_\infty)^\Delta$, and this latter group is zero because Δ acts non-trivially on D; here we have identified Δ with $G(L_\infty/\mathbb{Q}_\infty)$. As is explained in the proof of Theorem 5.2 in [C-G], we have a natural surjection

$$H^1(\mathbb{Q}_7, B_1) \to H^1(\mathbb{Q}_7, D). \tag{60}$$

By Tate duality, the universal norm subgroup will be the whole of $B_1(\mathbb{Q}_7)$ provided we can show that (60) is an isomorphism. As B_1 has good reduction over L, it is explained in [C-G, 5.31] that we have an exact sequence

$$0 \longrightarrow \tilde{E}(\mathbb{F}_7) \longrightarrow H^1(L, B_1) \overset{\phi}{\longrightarrow} H^1(L, D) \longrightarrow 0, \qquad (61)$$

where $\tilde{E}$ denotes the curve (50) over $\mathbb{F}_7$. Recall that Δ acts non-trivially on $\tilde{E}(\mathbb{F}_7)$. We may identify (60) with the map induced by ϕ in (61) between the Δ-invariants, as Δ is of order prime to 7. This shows that (60) is an isomorphism. This completes the proof of Theorem 4.11.

$\square$

4.12. Theorem. *We have* $\mathcal{S}(B_2/\mathbb{Q}_\infty) = 0$.

Proof. We again have $S = \{2, 3, 7\}$ and the fundamental diagram

$$
\begin{array}{ccccccc}
0 & \longrightarrow & \mathcal{S}(B_2/\mathbb{Q}_\infty)^\Gamma & \longrightarrow & H^1(G_{S,\infty}, B_{2,7\infty})^\Gamma & \overset{\phi_\infty}{\longrightarrow} & \underset{v \in S}{\oplus} J_v(\mathbb{Q}_\infty)^\Gamma \\
& & \big\uparrow{\alpha} & & \big\uparrow{\beta} & & \big\uparrow{\gamma} \\
0 & \longrightarrow & \mathcal{S}(B_2/\mathbb{Q}) & \longrightarrow & H^1(G_S, B_{2,7\infty}) & \overset{\lambda}{\longrightarrow} & \underset{v \in S}{\oplus} J_v(\mathbb{Q}).
\end{array}
$$

By Proposition 1.9 of Chapter 1, Coker λ can be identified with the dual of $B_2(\mathbb{Q})(7)$, and so has order 7. Also, Ker β has the same order as $B_2(\mathbb{Q})(7)$, and so it has order 7. Again β is surjective, and $\mathcal{S}(B_2/\mathbb{Q}) = 0$ by Fisher's calculations [F]. We conclude from the above diagram that we have the exact sequence

$$0 \longrightarrow \text{Ker } \beta \longrightarrow \text{Im } \lambda \cap \text{Ker } \gamma \longrightarrow \mathcal{S}(B_2/\mathbb{Q})^\Gamma \longrightarrow 0. \tag{62}$$

We claim that Ker γ has order 7^2. Indeed, both Ker γ_2 and Ker γ_3 have order 7 because $c_2 = c_3 = 7$ for B_2. Exactly as in the proof of Theorem 4.11, but working with the curve B_2 rather than B_1, one can show that γ_7 is injective. We omit the details.

We next show that

$$\#(\text{Im } \lambda \cap \text{Ker } \gamma) = 7. \tag{63}$$

Note that, in view of (62), this will imply that $\mathcal{S}(B_2/\mathbb{Q}_\infty)^\Gamma = 0$, and so $\mathcal{S}(B_2/\mathbb{Q}_\infty) = 0$. To prove (63), it suffices to show Ker γ maps onto

Coker $\lambda = \widehat{B_2(\mathbb{Q})}(7)$. To lighten the notation, put $B = B_2$ for the rest of the proof. By Tate duality, this latter assertion is equivalent to showing that the map

$$B(\mathbb{Q})(7) \to \bigoplus_{v \in S} B(\mathbb{Q}_v)/B_U(\mathbb{Q}_{\infty,v}) \tag{64}$$

is injective, where $B_U(\mathbb{Q}_{\infty,v})$ denotes the group of universal norms in $B(\mathbb{Q}_v)$ from $\mathbb{Q}_{\infty,v}$. We shall now show using the Tate parametrization of B over $\mathbb{Q}_2$ (see [Ta2] or [Si2] and the remark in Appendix A.3) that the universal norm subgroup $B_U(\mathbb{Q}_{\infty,2})$ contains no 7-torsion. Hence it cannot contain $B(\mathbb{Q})(7)$ and so the map in (64) is indeed injective, as required.

Let q denote the 2-adic Tate period of B over $\mathbb{Q}_2$. Then we have an analytic isomorphism

$$\xi : \mathbb{Q}_2^*/q^{\mathbb{Z}} \to B(\mathbb{Q}_2)$$

and it is well-known (see [Si2], p.431) that $\xi(\mathbb{Z}_2^*) = B_0(\mathbb{Q}_2)$, the subgroup of points of $B(\mathbb{Q}_2)$ with non-singular reduction. We claim that $\xi(\mathbb{Z}_2^*) = B_U(\mathbb{Q}_{\infty,2})$. Indeed, as $c_2 = 7$, it follows that $\xi(\mathbb{Z}_2^*)$ has index 7 in $B(\mathbb{Q}_2)$. But we also know that $B_U(\mathbb{Q}_{\infty,2})$ has index 7 in $B(\mathbb{Q}_2)$ (cf. [Lemma 3.4]). Moreover, as $\mathbb{Q}_{\infty,2}$ is the unramified $\mathbb{Z}_7$-extension of $\mathbb{Q}_2$, every element of $\mathbb{Z}_2^*$ is the norm of a unit in every finite extension of $\mathbb{Q}_2$ contained in $\mathbb{Q}_{\infty,2}$. Thus, by the Tate parametrisation, $\xi(\mathbb{Z}_2^*)$ is contained in $B_U(\mathbb{Q}_{\infty,2})$ and therefore these two groups must be equal. This completes the proof of Theorem 4.12.

$$\square$$

Chapter 5

Numerical examples over $\mathbb{Q}(\mu_{\mathrm{p}^\infty})$

INTRODUCTION

5.1. There is considerable interest in studying the Iwasawa theory of elliptic curves over the field $\mathbb{Q}(\mu_{p^\infty})$, for at least two different reasons. Firstly, as R. Greenberg pointed out to the first author seven or eight years ago, one gets a whole variety of phenomena occurring for the Selmer groups over $\mathbb{Q}(\mu_{p^\infty})$, which simply do not occur over the cyclotomic $\mathbb{Z}_p$-extension of $\mathbb{Q}$ (for example, the existence of finite non-zero Γ-submodules in the dual of the Selmer group over $\mathbb{Q}(\mu_{p^\infty})$). Secondly, as was already mentioned in **2.18**, there exist elliptic curves E over $\mathbb{Q}$ without complex multiplication for which $\mathbb{Q}(E_{p^\infty})$ is a pro-p extension of $\mathbb{Q}(\mu_p)$. For such curves E, we can often exploit knowledge of the Iwasawa theory of E over the field $\mathbb{Q}(\mu_{p^\infty})$ to prove deep statements, such as Conjectures 2.16 and 3.15, for the GL_2-Iwasawa theory of E over $\mathbb{Q}(E_{p^\infty})$ (see [C], [C-H]). We believe that much further interesting work can be done in this direction, both algebraically and analytically, and this is why we discuss some numerical examples in detail in this chapter. Most of the chapter is devoted to studying the isogeny class of elliptic curves of conductor 11 over the field $\mathbb{Q}(\mu_{5^\infty})$. The results we establish were communicated to the first author orally by R. Greenberg, and they are stated without proof in [G1], and partly proven in [G2]. We give here somewhat different proofs. We also discuss the Iwasawa theory of the isogeny class consisting of the two elliptic curves B_1 and B_2 of conductor 294 (see **4.3**) over the field $\mathbb{Q}(\mu_{7^\infty})$, using descent data about B_1 over $\mathbb{Q}(\mu_7)$, which has been proved by T. Fisher [F].

General strategy for the curves of conductor 11 over $\mathbb{Q}(\mu_{5^\infty})$

5.2. Again, we let A denote any of the three curves A_0, A_1, A_2 of conductor 11. We first outline the general strategy that we will use to study the curves of conductor 11 over $\mathbb{Q}(\mu_{5^\infty})$. Put

$$F = \mathbb{Q}(\mu_5), \ K_\infty = \mathbb{Q}(\mu_{5^\infty}), \ \Gamma = G(K_\infty/F).$$

We shall show later by arguments of classical descent theory, that we have $\mathcal{S}(A_1/F) = 0$. This, in turn, implies that both $\mathcal{S}(A_0/F)$ and $\mathcal{S}(A_2/F)$ are finite. In fact, we shall show later that $\mathcal{S}(A_0/F) = 0$ and $\mathcal{S}(A_2/F)$ has order 5^4. We assume throughout that the set S consists of the primes of F that lie above 5 and 11. Now A/F satisifes the hypotheses of Theorem 3.3 and so from Proposition 3.9, we have an exact sequence

$$0 \longrightarrow \mathcal{S}(A/K_\infty) \longrightarrow H^1(G_{S,\infty}, A_{p^\infty}) \overset{\lambda_\infty}{\longrightarrow} \underset{v \in S}{\oplus} J_v(K_\infty) \longrightarrow 0. \tag{65}$$

Taking Γ-invariants along this sequence, we obtain the exact sequence

$$H^1(G_{S,\infty}, A_{p^\infty})^\Gamma \overset{\phi_\infty}{\longrightarrow} \underset{v \in S}{\oplus} J_v(K_\infty)^\Gamma \longrightarrow H^1(\Gamma, \mathcal{S}(A/K_\infty)) \longrightarrow 0 \tag{66}$$

since (cf. (33))

$$H^1(\Gamma, H^1(G_{S,\infty}, A_{p^\infty})) = 0.$$

Thus it is clear that in order to compute $H^1(\Gamma, \mathcal{S}(A/K_\infty))$, we need to analyse Coker ϕ_∞. We do this as before, by considering the following commutative square that is a part of the fundamental diagram (9):

$$\begin{array}{ccc} H^1(G_{S,\infty}, A_{p^\infty})^\Gamma & \overset{\phi_\infty}{\longrightarrow} & \underset{v \in S}{\oplus} J_v(K_\infty)^\Gamma \\ \beta \uparrow & & \gamma \uparrow \\ H^1(G_S, A_{p^\infty}) & \overset{\lambda}{\longrightarrow} & \underset{v \in S}{\oplus} J_v(F). \end{array} \tag{67}$$

We recall that the maps β and γ are surjective and since $\mathcal{S}(A/F)$ is finite, we know by Proposition 1.9 that

$$\mathrm{Coker} \ \lambda \simeq \widehat{A(F)}(p).$$

Clearly,

$$\text{Coker } \phi_\infty = \text{Coker}(\phi_\infty \circ \beta) = \text{Coker}(\gamma \circ \lambda).$$

Let

$$\tau \; : \; A(F)(p) \; \to \; \underset{v \in S}{\oplus} A(F_v)/A_U(K_{\infty,v})$$

be the natural composite homomorphism

$$A(F)(p) \hookrightarrow \underset{v \in S}{\oplus} A(F_v) \twoheadrightarrow \underset{v \in S}{\oplus} A(F_v)/A_U(K_{\infty,v})$$

where $A_U(K_{\infty,v})$ is the subgroup of universal norms, i.e.,

$$A_U(K_{\infty,v}) = \underset{K'}{\cap} N_{K'/F_v}(E(K')),$$

as K' varies over all finite extensions of F_v contained in $K_{\infty,v}$ and N_{K'/F_v} denotes the norm map. Clearly, from (67), we have an exact sequence

$$\text{Ker } \gamma \xrightarrow{\;\widehat{\tau}\;} \text{Coker } \lambda \longrightarrow \text{Coker}(\gamma \circ \lambda) \longrightarrow 0,$$

and we claim that the arrow on the left is the dual of τ. In particular, we get an isomorphism

$$\text{Coker } \widehat{\tau} \simeq \text{Coker } \phi_\infty. \tag{68}$$

We now justify the above claim. Let $\Gamma_v := G(K_{\infty,v}/F_v)$ for $v \in S$, where $K_{\infty,v}$ is as before. We have (cf. **3.13**)

$$\text{Ker } \gamma \simeq \underset{v \in S}{\oplus} H^1(\Gamma_v, A)(p).$$

The group $H^1(\Gamma_v, A)(p)$ is dual to $\underset{v \in S}{\oplus} A(F_v)/A_U(K_{\infty,v})$, (cf. **3.13**). Using the above remarks on the dual groups, it is clear that the claim is justified. Observe therefore that (68) can be re-written as

$$\widehat{\text{Ker } \tau} \simeq \text{Coker } \phi_\infty. \tag{69}$$

Thus our strategy in computing Coker ϕ_∞ will be to analyse the homomorphism τ.

5.3. To help the reader through the computations that follow, we present in this paragraph a preview of the results and the methods used. Because of (69), we focus on the study of the homomorphism τ. We have

$$A_0(F)_5 \;\simeq\; \mathbb{Z}/5 \oplus \mu_5, \quad A_1(F)_5 \;\simeq\; \mathbb{Z}/5, \quad A_2(F)_5 \;\simeq\; \mu_5.$$

The above assertions hold for A_1 and A_2 because the exact sequences (47) and (48) do not split as $G(\overline{\mathbb{Q}}/\mathbb{Q})$-modules.

It is very easy to see (cf. Lemma 5.6) that the subgroup of points corresponding to $\mathbb{Z}/5$ of $A_1(F)_5$ and $A_0(F)_5$ do not lie in the universal norm subgroup for the unique prime in F above 5, and therefore do not lie in Ker τ. For the points μ_5 of $A_0(F)_5$ and $A_2(F)_5$, the situation is more subtle, and we prove that in both cases μ_5 lies in the universal norm subgroup at the unique prime above 5. However, in the case of A_0, we show that μ_5 does not belong to the universal norm subgroup for each of the four primes v dividing 11. Hence μ_5 too does not lie in Ker τ and therefore Ker τ is trivial. For the curve $A_2(F)$, and the four primes v dividing 11, the quotient group $A_2(F_v)/A_{2,U}(F_v)$ is itself trivial, and hence in this case Ker τ is of order 5. These remarks determine Ker τ for all three curves, and in view of (69) and (66), we get

$$H^1(\Gamma, \mathcal{S}(A_i/K_\infty)) = \begin{cases} 0 & \text{if } i = 0 \\ 0 & \text{if } i = 1 \\ \mathbb{Z}/5 & \text{if } i = 2. \end{cases}$$

In fact, more is true for the curve A_1, as the following theorem shows.

5.4. Theorem. *We have*

$$\mathcal{S}(A_1/K_\infty) = 0$$

where $K_\infty = \mathbb{Q}(\mu_{5^\infty})$.

Proof. We begin by giving a direct proof of the fact that

$$H^1(\Gamma, \mathcal{S}(A_1/K_\infty)) = 0$$

using Theorem 3.11. Indeed, all the hypotheses of this theorem hold with $F = \mathbb{Q}(\mu_5)$ and $p = 5$, since $A_1(F)(5) \simeq \mathbb{Z}/5$ is not contained in the kernel of reduction modulo 5. Further, by Theorem 3.3, the Γ-Euler characteristic of $\mathcal{S}(A_1/K_\infty)$ is finite and given by

$$\chi(\Gamma, \mathcal{S}(A_1/K_\infty)) = \rho_5(A_1/F).$$

But $\rho_5(A_1/F) = 1$ since $c_v(A_1) = 1$ for each prime v of F above 11, and

$$A_1(F)(5) = 5, \quad \tilde{A}_1(\mathbb{F}_5) = \mathbb{Z}/5, \quad \text{Ш}(A_1/F)(5) = 0;$$

the last statement is true because $\mathcal{S}(A_1/F) = 0$ (cf. Theorem 5.18 below). Hence $H^0(\Gamma, \mathcal{S}(A_1/K_\infty)) = 0$ which implies that $\mathcal{S}(A_1/K_\infty) = 0$, thereby proving the theorem.

$\square$

We remark that this example shows that Theorem 3.11 is sometimes applicable in cases where Proposition 4.15 of [G2] cannot be used. However, Theorem 3.11 cannot be used for the curves A_0 and A_2 over F, and this is why we must make a more detailed study of the universal norms in these cases.

THE CURVE A_0

5.5. We now start on the computations for the curve A_0 and the cyclotomic $\mathbb{Z}_5$-extension $K_\infty = \mathbb{Q}(\mu_{5^\infty})$ over $F = \mathbb{Q}(\mu_5)$. In order to simplify notation, we shall denote the curve A_0 by X for the whole of this section. All other notation is as in the previous sections. Let $K_{\infty,5} = \mathbb{Q}_5(\mu_{5^\infty})$ and let $X_U(K_{\infty,5}) \subseteq X(F_5)$ be the universal norm subgroup where F_5 denotes the completion of F at the unique prime above 5. We shall study the homomorphism τ (cf. **5.2**) componentwise. Let τ_5 be the homomorphism

$$\tau_5 \; : \; X(F)_5 \; \to \; X(F_5)/X_U(K_{\infty,5})$$

which is the component of τ for the unique prime above 5.

5.6. Lemma. *The map τ_5 restricts to an injective map on the subgroup* $\mathbb{Z}/5 \subseteq X(F)_5$.

Proof. We use essentially the same arguments as in the proof of Theorem 3.11. Regarding $\mathbb{Z}/5$ as lying in $X(F_5)$, it suffices to show that every non-zero point P in $\mathbb{Z}/5$ does not belong to $X_U(K_{\infty,5})$. Take any non-zero point P in $\mathbb{Z}/5$. We will show that it is not a norm even from $F_{1,5} := \mathbb{Q}_5(\mu_{5^2})$. Consider the commutative diagram

$$
\begin{array}{ccc}
X(F_{1,5}) & \longrightarrow & \tilde{X}(\mathbb{F}_5) \\
{\scriptstyle N_{F_{1,5}/F_5}}\big\downarrow & & \big\downarrow{\scriptstyle 5} \\
X(F_5) & \longrightarrow & \tilde{X}(\mathbb{F}_5)
\end{array}
\qquad (70)
$$

where the left vertical arrow is the norm map and the right one is its reduction modulo 5 and the horizontal arrows are the reduction maps. If a non-zero point P of $\mathbb{Z}/5$ is a norm from $F_{1,5}$, it is clear from (70) that the reduction of P modulo 5 must be zero. But P cannot be in the kernel of reduction modulo 5, since it is defined over $\mathbb{Q}_5$. Hence the lemma is proved.

$$\square$$

5.7. We now study the image of the subgroup $\mu_5 \subset X(F_5)_5$ under τ_5. We show that this subgroup in fact does lie in the universal norm subgroup, or equivalently that $\tau_5(\mu_5) = 0$. The proof of this proceeds as follows. We first prove that a result about the divisible subgroup $H^1(F_5, \tilde{A}_{5\infty})_{\mathrm{div}}$ of $H^1(F_5, \tilde{A}_{5\infty})$ is equivalent to the assertion $\tau_5(\mu_5) = 0$. The first step is the following general proposition. Let p be any prime, L a finite extension of $\mathbb{Q}_p$ and $\mathfrak{M}_L$ the maximal ideal of the ring of integers of L. Let E be an elliptic curve over L, and write $\hat{E}$, $\tilde{E}$ for the formal group and reduction of E respectively.

5.8. Proposition. *Assume that E/L has good ordinary reduction and that $\hat{E}(\mathfrak{M}_L)(p)$ is of order p. Let L_∞ be the cyclotomic $\mathbb{Z}_p$-extension of L. Then the following statements are equivalent:*

(i) $\mathrm{Ker}\ \rho \subseteq H^1(L, \tilde{E}_{p^\infty})_{\mathrm{div}}$, *where ρ is the restriction map*

$$\rho\ :\ H^1(L, \tilde{E}_{p^\infty}) \to H^1(L_\infty, \tilde{E}_{p^\infty}).$$

(ii) *Every element of $\hat{E}(\mathfrak{M}_L)(p)$ is a universal norm from L_∞.*

Proof. Let $E_U^*(L_\infty)$ be the maximal pro-p subgroup of $E_U(L_\infty)$ (the former group is the same as the p-adic completion of the latter group). Then assertion (ii) is equivalent to the following statement:

$$E_U^*(L_\infty) \text{ has non-trivial torsion.} \tag{71}$$

Clearly $E_U^*(L_\infty)$ has non-trivial torsion if it contains $\hat{E}(\mathfrak{M}_L)(p)$. Conversely, suppose that $E_U^*(L_\infty)$ has non-trivial torsion. Then, a norm argument similar to that in **(3.13)** shows that any element of $E(L)(p)$ which is in $E_U(L_\infty)$ is necessarily in $\hat{E}(\mathfrak{M}_L)(p)$. Therefore we have to prove that (i) is equivalent to (71). Consider the Pontryagin dual $\widehat{E_U^*(L_\infty)}$ of $E_U^*(L_\infty)$. By duality, (71) is equivalent to the statement that $\widehat{E_U^*(L_\infty)}$ is not divisible. We now appeal to one of the main results of [C-G] (cf. [C-G, Theorem 5.2]), which gives an explicit description of the dual of $E_U^*(L_\infty)$. Under the Tate pairing, the dual of $E_U^*(L_\infty)$ is the group $H^1(L, \tilde{E}_{p^\infty})/\operatorname{Ker} \rho$ (cf. Appendix A.2.5). It is immediate now that (71) is equivalent to the assertion

$$H^1(L, \tilde{E}_{p^\infty})/\operatorname{Ker} \rho \text{ is not divisible.} \tag{72}$$

Since $\hat{E}(\mathfrak{M}_L)(p)$ is assumed to be of order p, it is easily seen that (72) is equivalent to the assertion (i) that $\operatorname{Ker} \rho \subseteq H^1(L, \tilde{E}_{p^\infty})_{\mathrm{div}}$. This completes the proof of the proposition.

$$\square$$

5.9. In this paragraph, we formulate the equivalent statements of Proposition 5.8 in yet another way for the curve $X = A_0$ over $F_5 = \mathbb{Q}_5(\mu_5)$. Let $\mathfrak{M}_5$ be the maximal ideal of the ring of integers of F_5. Then $\hat{X}(\mathfrak{M}_5)(5)$ is of exact order 5 because it cannot contain a 5^2-division point of $\hat{X}$. Indeed, a point of order 5^2 on $\hat{X}$ can be rational only over an extension of $\mathbb{Q}_5$ of ramification index 20 (cf. [Si], Chapter IV, §6). Hence by Proposition 5.8, in order to prove that $\tau_5(\mu_5) = 0$, it suffices to show that $\operatorname{Ker} \rho \subseteq H^1(F_5, \tilde{X}_{5\infty})_{\mathrm{div}}$ where

$$\rho : H^1(F_5, \tilde{X}_{5\infty}) \to H^1(K_{\infty,5}, \tilde{X}_{5\infty})$$

is the restriction map as before. Now Tate duality gives a dual pairing

$$H^1(F_5, T_5(\hat{X}_{5\infty})) \times H^1(F_5, \tilde{X}_{5\infty}) \to \mathbb{Q}_5/\mathbb{Z}_5. \tag{73}$$

Indeed, since the Weil pairing restricts to a non-degenerate pairing

$$\hat{X}_{5^n} \times \tilde{X}_{5^n} \to \mu_{5^n},$$

the usual cup-product

$$H^1(F_5, \hat{X}_{5^n}) \times H^1(F_5, \tilde{X}_{5^n}) \to H^2(F_5, \mu_{5^n}) \simeq \mathbb{Z}/5^n \tag{74}$$

is a non-degenerate pairing. The pairing in (73) in then obtained from the above pairing on taking limits. The exact orthogonal complement of $H^1(F_5, \tilde{X}_{5\infty})_{\mathrm{div}}$ in (73) is the torsion subgroup $H^1(F_5, T_5(\hat{X}_{5\infty}))_{\mathrm{tors}}$. This group is easily computed using Kummer theory on $\hat{X}$. There is an exact sequence

$$0 \longrightarrow \hat{X}(\mathfrak{M}_5) \longrightarrow H^1(F_5, T_5(\hat{X}_{5\infty})) \longrightarrow T_5(H^1(F_5, \hat{X})) \longrightarrow 0.$$

The group $T_5(H^1(F_5, \hat{X}))$ is zero as $H^1(F_5, \hat{X})$ is a finite group (by Lemma 3.6) and hence

$$H^1(F_5, T_5(\hat{X}_{5^\infty}))_{\text{tors}} \simeq \hat{X}(\mathfrak{M}_5)(5).$$

It emerges from the above discussion that in order to show that $\tau_5(\mu_5)$ is zero, it suffices to show that the group $\hat{X}(\mathfrak{M}_5)(5)$ is orthogonal to Ker $\rho \subseteq H^1(F_5, \tilde{X}_{5^\infty})_{\text{div}}$ in (73). But

$$\text{Ker } \rho \simeq H^1(\Gamma_5, \tilde{X}_5) = \text{Hom}(\Gamma_5, \tilde{X}_5) \simeq \mathbb{Z}/5.$$

We have therefore shown

$$\tau_5(\mu_5) = 0 \iff \begin{array}{c} \mu_5 = \hat{X}(\mathfrak{M}_5)(5) \text{ is orthogonal to } H^1(\Gamma_5, \tilde{A}_5) \\ \text{under the Tate local duality pairing.} \end{array} \tag{75}$$

5.10. We next discuss how the right hand side of (75) follows from a cup-product calculation. Let α be a generator of $\mu_5 \simeq \hat{X}_5(\mathfrak{M}_5)$. Consider the composite

$$\hat{X}_5(\mathfrak{M}_5) \to \hat{X}(\mathfrak{M}_5)/5 \hookrightarrow H^1(F_5, \hat{X}_5)$$

where the first map is the canonical one and the second map is part of the Kummer sequence for $\hat{X}$. Denote by $\tilde{\alpha}$ the image of α in $H^1(F_5, \hat{X}_5)$ under this composite map. Similarly, if β is a generator of

$$\mathbb{Z}/5 \simeq H^1(\Gamma_5, \tilde{X}_5) \hookrightarrow H^1(F_5, \tilde{X}_5),$$

let $\tilde{\beta}$ denote its image in $H^1(F_5, \tilde{X}_5)$. Then the pairing (α, β) under the Tate local duality pairing is just the cup-product $(\tilde{\alpha}, \tilde{\beta})$ in (74), which in turn is the restriction of the cup-product

$$H^1(F_5, X_5) \times H^1(F_5, X_5) \to H^2(F_5, \mu_5) \simeq \mathbb{Z}/5$$

induced by the Weil pairing. Since the group X_5 is rational over F_5, we have $H^1(F_5, X_5) = \text{Hom}(G(\overline{F}_5/F_5), X_5)$. We carry out this cup-product calculation now, using well-known folklore relating the Weil pairing and multiplicative Kummer generators which is explained in general in Appendix A.4. Let $F_{1,5} = \mathbb{Q}_5(\mu_{5^2})$. Since Γ_5 has a unique quotient of order 5, which is $G(F_{1,5}/F_5)$, we have

$$H^1(\Gamma_5, \tilde{X}_5) = \text{Hom}(G(F_{1,5}/F_5), \tilde{X}_5).$$

The extension determined by $\tilde{\alpha}$ is $F_5(\hat{X}_{5^2})/F_5$ and that determined by $\tilde{\beta}$ is $F_5(\mu_{5^2})/F_5$ (cf. Corollary A.4.2). Since $\mu_5 \subset F_5$, let ξ be such that $F_5(\hat{X}_{5^2}) = F_5(\xi^{1/5})$. Let ζ be a primitive 5th-root of unity. Then by Lemma A.4.4, we see that the Tate pairing $(\tilde{\alpha}, \tilde{\beta}) = 0$ if and only if the 5-Hilbert norm residue symbol $(\xi, \zeta)_5 = 1$. Summarizing, we now have

$$\tau_5(\mu_5) = 0 \iff (\xi, \zeta)_5 = 1.$$

5.11. Proposition. *The 5-Hilbert norm-residue symbol* $(\xi, \zeta)_5 = 1$.

Proof. Recall that the Hilbert symbol $(\xi, \zeta)_5 = 1$ if ξ is a norm from $F_{1,5} = F_5(\zeta^{1/5})$. Write $(\xi, F_{1,5}/F_5)$ for the local Artin symbol of ξ in $G(F_{1,5}/F_5)$. Hence we must show that

$$(\xi, F_{1,5}/F_5) = 1.$$

But by the basic functoriality of the Artin symbol, we have

$$(\xi, F_{1,5}/F_5) = (\eta, F_{1,5}/\mathbb{Q}_5), \tag{76}$$

where $\eta = N_{F_5/\mathbb{Q}_5}(\xi)$, and the right hand side of (76) denotes the Artin symbol for the abelian extension $F_{1,5}/\mathbb{Q}_5$. We now show that $\eta \in \mathbb{Q}_5^{*^5}$, which will clearly prove the proposition. Write $\Delta = G(F_5/\mathbb{Q}_5)$ and

$$\omega \; : \; \Delta \to (\mathbb{Z}/5\mathbb{Z})^*$$

be the character giving the action of Δ on μ_5. The crucial point is that the extension $F_5(\hat{X}_{5^2})$ is also abelian over $\mathbb{Q}_5$. Hence $\Delta = G(F_5/\mathbb{Q}_5)$ acts trivially on $G(F_5(\hat{X}_{5^2})/F_5)$ via inner automorphisms. But multiplicative Kummer theory gives a Δ-isomorphism

$$\xi F_5^{*^5}/F_5^{*^5} \simeq \mathrm{Hom}(G(F_5(\hat{X}_{5^2})/F_5), \mu_5).$$

Thus, we must have that Δ acts on $\xi F_5^{*^5}/F_5^{*^5}$ via ω, where ω is as above. Hence $N_{F_5/\mathbb{Q}_5}(\xi) \in \mathbb{Q}_5^{*^5}$ because ω is a non-trivial character of Δ and the proposition is proved.

$\square$

We have thus finally established the following result.

5.12. Corollary. *The subgroup* $\mu_5 \subset X(F)_5$ *lies in the universal norm subgroup* $X_U(K_{\infty,5}) \subseteq X(F_5)$.

$\square$

5.13. Let $K_{n,5}$ denote the field $\mathbb{Q}_5(\mu_{5^{n+1}})$. Corollary 5.12 shows that, given any $R \in \mu_5 \subset X(F_5)$, then for each $n \geq 1$ there exists $R_n \in X(K_{n,5})$ such that $N_{K_{n,5}/F_5}(R_n) = R$. We remark that in fact we can choose R_n to lie on the formal group $\hat{X}(\mathfrak{M}_{n,5})$, where $\mathfrak{M}_{n,5}$ denotes the maximal ideal of the ring of integers of $K_{n,5}$. This is because we can clearly replace R_n by any point of the form $R_n + W_n$, where W_n belongs to $\mathbb{Z}/5 \subset X(F_5)$. Since the subgroup $\mathbb{Z}/5 \subset X(F_5)$ maps surjectively onto the reduction $\tilde{X}(\mathbb{Z}/5)$ of X modulo 5, we simply choose W_n so that $R_n + W_n$ reduces to zero modulo 5, and therefore lies in $\hat{X}(\mathfrak{M}_{n,5})$.

5.14. We turn our attention now to the primes $v \in S$ that lie above 11. The prime 11 splits completely in F and it suffices to consider any one of the four primes that lie above 11. Fix one such prime v. Then $F_v = \mathbb{Q}_{11}$ and $K_{\infty,v} = \mathbb{Q}_{11}(\mu_{5^\infty})$ is the unique unramified $\mathbb{Z}_5$-extension of $\mathbb{Q}_{11}$. Recall that we are interested in computing Ker τ. Also recall from Lemma 5.6 that τ is injective on the subgroup $\mathbb{Z}/5$ of $X(F)(5)$. By Corollary 5.12, the subgroup μ_5 maps into the universal norm subgroup $X_U(K_{\infty,5})$. However, we will now show that for v dividing 11, the v-component of τ

$$\tau_v : X(F)(5) \;\to\; X(\mathbb{Q}_{11})/X_U(K_{\infty,v})$$

is injective.

5.15. Proposition. *The subgroup $\mu_5 \subset X(F)(5)$ does not lie in the universal norm subgroup $X_U(K_{\infty,v})$, i.e., $\tau_v(\mu_5) \neq 0$, for v dividing 11.*

Proof. Let v be any prime of F dividing 11. Since X has split multiplicative reduction at v, it admits an 11-adic Tate parametrization over $\mathbb{Q}_{11}$ (see [Ta2] or [Si2], and the remarks in Appendix A.3). Let q_0 denote the 11-adic Tate period of X. Then we have an analytic surjection

$$\xi : \overline{\mathbb{Q}}_{11}^* \to X(\overline{\mathbb{Q}}_{11}) \tag{77}$$

whose kernel is precisley $q_0^{\mathbb{Z}}$. As $K_{\infty,v}$ is the unique unramified $\mathbb{Z}_5$-extension of $\mathbb{Q}_{11}$, the index of $X_U(K_{\infty,v})$ in $X(\mathbb{Q}_{11})$, which is equal to the order of $H^1(\Gamma_v, X(K_{\infty,v}))$, is equal to $c_{11}^{(5)} = 5$ (see Lemma 3.4). Let $\mathbb{Z}_{11}^*$ denote the group of units of $\mathbb{Z}_{11}$. Since $K_{\infty,v}$ is the unramified $\mathbb{Z}_5$-extension of $\mathbb{Q}_{11}$, every element of $\mathbb{Z}_{11}^*$ is a norm from every finite extension of $\mathbb{Q}_{11}$ which is contained in $K_{\infty,v}$. Thus, as $\mathrm{ord}_{11}(q_0) = 5$, we must have

$$X_U(K_{\infty,v}) = \xi(\mathbb{Z}_{11}^*). \tag{78}$$

In order to avoid confusion between groups of points on X and subgroups of $\mathbb{Q}_{11}^*$, let us, for the rest of the proof, write W for the subgroup μ_5 of $X(F)$. In view of (78), the proposition will follow provided we can show

$$\xi(\mu_5) \neq W. \tag{79}$$

But this is clear from the isogeny

$$0 \to W \to X = A_0 \to A_1 \to 0.$$

Indeed, let q_1 be the 11-adic Tate period of A_1. By virtue of the existence of this isogeny, we must have $q_1^5 = q_0$ (see [Ta2], p. 344), and the isogeny is then given by the commutative diagram

$$
\begin{array}{ccccccccc}
& & 1 & \longmapsto & q_0 & & & & \\
0 & \longrightarrow & \mathbb{Z} & \longrightarrow & \mathbb{Q}_{11}^* & \longrightarrow & A_0(\mathbb{Q}_{11}) & \longrightarrow & 0 \\
& & \downarrow{\scriptstyle 5} & & \downarrow{\scriptstyle 1} & & \downarrow & & \\
0 & \longrightarrow & \mathbb{Z} & \longrightarrow & \mathbb{Q}_{11}^* & \longrightarrow & A_1(\mathbb{Q}_{11}) & \longrightarrow & 0. \\
& & 1 & \longmapsto & q_1 & & & &
\end{array}
$$

It is plain from this diagram that $\xi(\mu_5)$ is not equal to the kernel of the right hand vertical map, proving (79). This completes the proof of the proposition.

$\square$

Note that we have finally proved

5.16. Theorem. *For the curve A_0, the prime $p = 5$ we have*

$$H^1(\Gamma, \mathcal{S}(A_0/\mathbb{Q}(\mu_{5^\infty}))) = 0.$$

Proof. Recall that τ is a direct sum of τ_5 and the τ_v's, where v runs over the primes dividing 11. Suppose $\tau(z) = 0$, where $z = x + y$, with $x \in \mathbb{Z}/5$ and $y \in \mu_5$. But, by Corollary 5.12, $\tau_5(y) = 0$ and so $\tau_5(x) = 0$. On the other hand, by Lemma 5.6, τ_5 is injective on $\mathbb{Z}/5$ and so $x = 0$. Hence $\tau(y) = 0$ and so $\tau_v(y) = 0$, whence $y = 0$ by Proposition 5.15. Thus $\operatorname{Ker} \tau = 0$, but by **(5.2)**, (66), and (69),

$$\widehat{\operatorname{Ker} \tau} \simeq \operatorname{Coker} \phi_\infty \simeq H^1(\Gamma, \mathcal{S}(A_0/K_\infty)).$$

$$\text{THE CURVE } A_2$$

We can straightaway reap the benefits of the work done in the previous section now to study the curve A_2.

5.17. Theorem. *For the curve A_2 and the prime $p = 5$, we have*

$$H^1(\Gamma, \mathcal{S}(A_2/\mathbb{Q}(\mu_{5\infty}))) = \mathbb{Z}/5.$$

Proof. We have (cf. **5.3**)

$$A_2(F)(5) \simeq \mu_5.$$

Further, $A_2 = A_0/(\mathbb{Z}/5)$, and the isogeny from A_0 to A_2 induces an isomorphism from $\hat{A}_0$ to $\hat{A}_2$, where $\hat{A}_i$ denotes the formal group of A_i at the prime 5; this is clear because $\mathbb{Z}/5$ does not lie on $\hat{A}_0$. Now, as was remarked in **(5.13)**, each element of $\mu_5 \in \hat{A}_0(\mathfrak{M}_5)$ is a norm of a point in $\hat{A}_0(\mathfrak{M}_{n,5})$ for all $n \geq 0$, where $\mathfrak{M}_{n,5}$ is the maximal ideal of the ring of integers of $\mathbb{Q}_5(\mu_{5^{n+1}})$. Hence the same is true for μ_5 when viewed as a subgroup of the points on the isomorphic formal group $\hat{A}_2$. In particular, we see that the subgroup μ_5 of $A_2(F_5)$ lies in the universal norm subgroup $A_{2,U}(K_{\infty,5})$. However, for A_2, the points in μ_5 are also universal norms from $K_{\infty,v}$, for each prime v of F dividing 11. This is because $A_{2,U}(K_{\infty,v})$ is the whole of $A_2(\mathbb{Q}_{11})$ by Lemma 3.4, since $c_v = 1$ for each v dividing 11, as the discriminant of A_2 is -11. Thus $\tau(\mu_5) = 0$ and Ker τ is of order 5. Since $\widehat{\text{Ker } \tau} \simeq H^1(\Gamma, \mathcal{S}(A_2/K_\infty))$ as before, the theorem is proved.

$$\square$$

$$\text{INFINITE DESCENT ON } A_1 \text{ OVER } \mathbb{Q}(\mu_5)$$

The aim of this section is to use classical descent theory à la Mordell-Weil to prove the following result:

5.18. Theorem. *We have $\mathcal{S}(A_1/F) = 0$ when $F = \mathbb{Q}(\mu_5)$ and $p = 5$.*

5.19. As has already been explained (see the proof of Theorem 5.4), Theorem 5.18 implies easily that $\mathcal{S}(A_1/\mathbb{Q}(\mu_{5\infty})) = 0$. Our arguments to prove Theorem 5.18 have been inspired by those in §5 of Greenberg's article [G2], where it is shown that $\mathcal{S}(A_1/\mathbb{Q}) = 0$ for $p = 5$. However, more delicate arguments are required to carry out the descent over $\mathbb{Q}(\mu_5)$, and we have therefore taken the liberty of including rather full details.

We shall use the following notation throughout this section. Put $F = \mathbb{Q}(\mu_5)$. We recall that 5 is totally ramified and 11 splits completely in F, and we let S denote the set of primes of F above 5 or 11. As usual, F_S will denote the maximal extension of F unramified outside S and the archimedean primes of F, and we put $G_S = G(F_S/F)$ as always. It will be more convenient for us to work with the 5-Selmer group $\mathcal{S}(A_1/F, 5)$ which is defined by the exactness of the sequence

$$0 \longrightarrow \mathcal{S}(A_1/F, 5) \longrightarrow H^1(G_S, A_{1,5}) \longrightarrow \bigoplus_{v \in S} (H^1(F_v, A_1))_5.$$

Then, as is explained in the proof of Lemma 1.8, we have the exact sequence

$$0 \longrightarrow \Omega \longrightarrow \mathcal{S}(A_1/F, 5) \longrightarrow (\mathcal{S}(A_1/F))_5 \longrightarrow 0, \qquad (80)$$

where $\Omega = \mathbb{Z}/5\mathbb{Z}$ denotes the subgroup of $A_{1,5}$ which is generated by the rational point $(0,0)$ of order 5. Here we have used the fact that $A_{1,5^\infty}(F) = \Omega$. Hence Theorem 5.18 is equivalent to showing that $\Omega = \mathcal{S}(A_1/F, 5)$, or equivalently, that every element of $\mathcal{S}(A_1/F, 5)$ arises from dividing some multiple of $(0,0)$ by 5.

We analyse $\mathcal{S}(A_1/F, 5)$ via the exact sequence (47) of G_S-modules. To avoid possible confusion with subgroups of the multiplicative group, let us put

$$\Theta = A_{1,5}/\Omega.$$

Thus Θ is isomorphic to μ_5 as a Galois module, and we have the exact sequence of Galois modules (which is just (47) re-written in our new notation)

$$0 \longrightarrow \Omega \longrightarrow A_{1,5} \longrightarrow \Theta \longrightarrow 0. \qquad (81)$$

Now $A_{1,5}(F) = \Omega$ because (81) does not split, which in turn is because the discriminant of A_1 is -11. Hence, taking G_S-cohomology of (81), we obtain the exact sequence

$$0 \longrightarrow \Theta \longrightarrow H^1(G_S, \Omega) \xrightarrow{\;g\;} H^1(G_S, A_{1,5}) \xrightarrow{\;f\;} H^1(G_S, \Theta)$$
$$(82)$$

where f and g denote the induced maps as indicated. We recall that $\mathcal{S}(A_1/F, 5)$ is a subgroup of $H^1(G_S, A_{1,5})$.

5.20. Proposition. *We have an inclusion* $\mathcal{S}(A_1/F, 5) \subset \mathrm{Ker}\; f$.

Proof. Observe that since μ_5 is rational over F, we have $H^1(G_S, \Theta) = \mathrm{Hom}(G_S, \mu_5)$. The proof is immediate from the local Lemma 5.21 given below, and assertion (i) of Proposition 5.26, about the global arithmetic of F. Indeed, if α is any element of $\mathcal{S}(A_1/F, 5)$, let L_α be the fixed field of the kernel of $f(\alpha)$. By Lemma 5.21 below, L_α must be a cyclic extension of F, which is unramified outside 5, and in which every prime of F dividing 11 splits completely. Also, the degree of L_α over F must divide 5. By (i) of Proposition 5.26 below, there is no non-trivial extension of F with these properties. Hence $L_\alpha = F$, and so $f(\alpha) = 0$, completing the proof of the proposition.

$\square$

5.21. Lemma. *For $\alpha \in \mathcal{S}(A_1/F, 5)$, let L_α denote the fixed field of $f(\alpha)$. Then L_α is a cyclic extension of F of degree dividing 5, which is unramified outside the unique prime of F dividing 5, and in which all the four primes of F dividing 11 split completely.*

Proof. All is clear, except the final assertion. Let v be any prime of F dividing 11, and fix an identification of F_v with $\mathbb{Q}_{11}$. Let s_v denote the restriction map from $H^1(G_S, \Theta)$ to $H^1(\mathbb{Q}_{11}, \Theta)$. We must show that $s_v(f(\alpha)) = 0$ for all $\alpha \in \mathcal{S}(A_1/F, 5)$.

As A_1 is a Tate curve over $\mathbb{Q}_{11}$, we have the canonical exact sequence of $G(\overline{\mathbb{Q}}_{11}/\mathbb{Q}_{11})$-modules

$$0 \longrightarrow C_{11} \longrightarrow A_{1,5^\infty} \longrightarrow D_{11} \longrightarrow 0, \tag{83}$$

where C_{11} is isomorphic to μ_{5^∞} and D_{11} is isomorphic to $\mathbb{Q}_5/\mathbb{Z}_5$ (with trivial Galois action) as $G(\overline{\mathbb{Q}}_{11}/\mathbb{Q}_{11})$-modules. We claim that the global subgroup Ω of $A_1(\mathbb{Q}_{11})$ can be identified with $(C_{11})_5$ in the exact sequence (83). To prove this, we note that we have an isogeny φ

$$0 \longrightarrow \Omega \longrightarrow A_1 \overset{\varphi}{\longrightarrow} A_0 \longrightarrow 0.$$

Let q_0 be the 11-adic Tate period of A_0, and q_1 the 11-adic Tate period of A_1. As was already remarked in the proof of Proposition 5.15 we must have $q_0 = q_1^5$. Thus, at the level of Tate curves, the isogeny ϕ must be of the form (see [Ta2], p. 324)

$$
\begin{array}{ccccccccc}
& & & & 1 & \longmapsto & q_1 & & \\
0 & \longrightarrow & \mathbb{Z} & \longrightarrow & \mathbb{Q}_{11}^* & \longrightarrow & A_1(\mathbb{Q}_{11}) & \longrightarrow & 0 \\
& & \downarrow{\scriptstyle 1} & & \downarrow{\scriptstyle 5} & & \downarrow{\scriptstyle \varphi} & & \\
0 & \longrightarrow & \mathbb{Z} & \longrightarrow & \mathbb{Q}_{11}^* & \longrightarrow & A_0(\mathbb{Q}_{11}) & \longrightarrow & 0. \\
& & & & 1 & \longmapsto & q_0 & &
\end{array}
$$

It is clear from this diagram that $\Omega = \mathrm{Ker}\ \varphi = (C_{11})_5$. It follows from (83) that we can identify Θ with $(D_{11})_5$ as $G(\overline{\mathbb{Q}}_{11}/\mathbb{Q}_{11})$-modules. Now let α denote any element of $\mathcal{S}(A_1/F, 5)$. Let t_v be the composite

$$
t_v : H^1(G_S, \Theta) \xrightarrow{\ s_v\ } H^1(\mathbb{Q}_{11}, (D_{11})_5) \xrightarrow{\ u_v\ } H^1(\mathbb{Q}_{11}, D_{11}),
$$

where u_v is the map induced by the inclusion of $(D_{11})_5$ into D_{11}. Note that u_v is clearly injective, since $G(\overline{\mathbb{Q}}_{11}/\mathbb{Q}_{11})$ acts trivially on D_{11}. We claim that we always have $t_v(f(\alpha)) = 0$. Indeed, we have the commutative diagram

$$
\begin{array}{ccccc}
H^1(G_S, A_{1,5}) & \xrightarrow{\ h\ } & H^1(G_S, A_{1,5^\infty}) & \xrightarrow{\ r_v\ } & H^1(\mathbb{Q}_{11}, A_{1,5^\infty}) \\
\downarrow{\scriptstyle f} & & & & \downarrow \\
H^1(G_S, \Theta) & & \xrightarrow{\hspace{3cm} t_v \hspace{3cm}} & & H^1(\mathbb{Q}_{11}, D_{11})
\end{array}
$$

where h is the induced map, and r_v is the restriction map. But $h(\alpha)$ is in $\mathcal{S}(A_1/F)$, and so $r_v(h(\alpha))$ must be zero because it lies inside the subgroup $A_1(\mathbb{Q}_{11}) \otimes \mathbb{Q}_5/\mathbb{Z}_5 = 0$ of $H^1(\mathbb{Q}_{11}, A_{1,5^\infty})$. Thus it is clear from the diagram that $t_v(f(\alpha)) = 0$, whence $s_v(f(\alpha)) = 0$ because, as was remarked above, the map u_v is injective. This completes the proof of the lemma.

$\square$

The beauty of Proposition 5.20 is that it tells us that the image of g in the exact sequence (82) contains $\mathcal{S}(A_1/F, 5)$. We now proceed to study the properties of the extensions of F which arise from those β in $\mathrm{Hom}(G_S, \Omega)$ with $g(\beta)$ in $\mathcal{S}(A_1/F, 5)$. The first lemma we need is already established in §5 of [G2] as an important step in the proof given there that $\mathcal{S}(A_1/\mathbb{Q}) = 0$ for $p = 5$. Let $T = \{5, 11\}$. Clearly, F_S is also the maximal extension of $\mathbb{Q}$ unramified outside 5, 11 and ∞. Put $G_T = G(F_S/\mathbb{Q})$.

5.22. Lemma. *Let* $\mathbb{Q}(\mu_{11})^+$ *be the maximal real subfield of the field* $\mathbb{Q}(\mu_{11})$ *of 11-th roots of unity. Then the kernel of the natural map*

$$\eta \,:\, H^1(G_T, \Omega) \to H^1(G_T, A_{1,5\infty})$$

is precisely $\mathrm{Hom}(G(\mathbb{Q}(\mu_{11})^+/\mathbb{Q}), \Omega)$.

Proof. Let W denote the kernel of η. Now it is clear that

$$H^1(G_T, \Omega) = \mathrm{Hom}(G(R/\mathbb{Q}), \Omega)$$

where R is the compositum of $\mathbb{Q}(\mu_{11})^+$ and the first layer of the cyclotomic $\mathbb{Z}_5$-extension of $\mathbb{Q}$. Hence W has order at most 5^2. On the other hand, W is certainly non-trivial because it must contain the inverse image of the subgroup Ω of $H^1(G_T, A_{1,5})$ under the natural map η by Proposition 5.20 and (80). Hence the assertion of the lemma will follow if we can show that every element of W must factor through an extension of $\mathbb{Q}$ which is unramified at 5. Let $\tilde{A}_1$ denote the reduction of A_1 modulo 5, and let $\tilde{A}_{1,5\infty}$ denote the group of all 5-power division points on $\tilde{A}_1$. Let I_5 denote the inertial subgroup of some fixed prime of $\overline{\mathbb{Q}}$ above 5. We clearly have the commutative diagram

$$\begin{array}{ccccc}
H^1(G_T, \Omega) & \longrightarrow & H^1(G_T, A_{1,5\infty}) & \longrightarrow & H^1(\mathbb{Q}_S, \tilde{A}_{1,5\infty}) \qquad (84) \\
\downarrow & & & & \downarrow \\
H^1(I_5, \Omega) & \longrightarrow\! & & \longrightarrow & H^1(I_5, \tilde{A}_{1,5\infty}).
\end{array}$$

But Ω maps injectively under reduction modulo 5, and the bottom horizontal arrow is then injective because I_5 acts trivially on Ω and $\tilde{A}_{1,5\infty}$. Thus every element in W must map to zero in the left vertical arrow, and therefore the extensions of $\mathbb{Q}$ corresponding to elements of W must be unramified at 5, as required. This completes the proof of the lemma.

$$\square$$

We now analyse the local conditions which arise for those β in $\mathrm{Hom}(G_S, \Omega)$ such that $g(\beta) \in \mathcal{S}(A_1/F, 5)$. The next lemma covers the case of primes v of F dividing 11. Put $G_{11} = G(\overline{\mathbb{Q}}_{11}/\mathbb{Q}_{11})$. Taking G_{11}-invariants of the non-split exact sequence (81) of G_{11}-modules, we obtain the exact sequence

$$0 \longrightarrow \Theta \longrightarrow \mathrm{Hom}(G_{11}, \Omega) \xrightarrow{\;g_{11}\;} H^1(\mathbb{Q}_{11}, A_{1,5}). \qquad (85)$$

Put $Y_{11} = A_1(\mathbb{Q}_{11})/5$. Local Kummer theory on A_1 gives, as usual, the canonical injection $Y_{11} \hookrightarrow H^1(\mathbb{Q}_{11}, A_{1,5})$. If $\beta \in \mathrm{Hom}(G_{11}, \Omega)$, we write R_β for the fixed field of the kernel of β. As usual, we write $\mathbb{Q}_{11}(\mu_{11})^+ = \mathbb{Q}_{11}(\zeta_{11} + \zeta_{11}^{-1})$, where ζ_{11} is any primitive 11-th root of unity.

5.23. Lemma. *Let x be any element of the subgroup Y_{11} of $H^1(\mathbb{Q}_{11}, A_{1,5})$. Let M_x denote the compositum of all fields R_β for β running over all elements of $\mathrm{Hom}(G_{11}, \Omega)$ with $g_{11}(\beta) = x$. If $x \neq 0$, then M_x is an unramified extension of $\mathbb{Q}_{11}(\mu_{11})^+$. If $x = 0$, then M_x is contained in an unramified extension of $\mathbb{Q}_{11}(\mu_{11})^+$.*

Proof. Let $\xi : \overline{\mathbb{Q}}_{11}^* \to A_1(\mathbb{Q}_{11})$ denote the Tate parametrization. We have already shown in the proof of Lemma 5.21 that $\xi(\mu_5) = \Omega$. It follows easily that the image of Θ is $\mathrm{Hom}(G_{11}, \Omega)$ in (85) is generated by the homomorphism $\sigma \mapsto \xi(\sigma(q_1^{1/5}))/q_1^{1/5}$, where $q_1^{1/5}$ is any fixed 5-th root of the Tate period q_1 of A_1. It is plain that the fixed field of the kernel of this homomorphism is $\mathbb{Q}_{11}(q_1^{1/5})$. On the other hand, Lemma 5.22 shows that the natural map from $\mathrm{Hom}(G_T, \Omega)$ to $H^1(G_T, A_{1,5})$ sends $\mathrm{Hom}(G(\mathbb{Q}(\mu_{11})^+/\mathbb{Q}), \Omega)$ to the subgroup Ω of $H^1(G_T, A_{1,5})$. Write $H = \mathbb{Q}_{11}(\mu_{11})^+$. It follows immediately from this last fact that

$$g_{11}(\mathrm{Hom}(G(H/\mathbb{Q}_{11}), \Omega)) \subset Y_{11}. \tag{86}$$

But $\# Y_{11} = 5$ because $c_{11}(A_1) = 1$, and thus we see that the inclusion in (86) must in fact be an equality because the group on the left of (86) is not zero. Hence $M_x = H(q_1^{1/5})$ if $x \neq 0$ and $M_x = \mathbb{Q}_{11}(q_1^{1/5})$ if $x = 0$. But q_1 is a local parameter of $\mathbb{Q}_{11}$, and both H and $\mathbb{Q}_{11}(q_1^{1/5})$ are tamely ramified cyclic extensions of $\mathbb{Q}_{11}$. Let $\mathbb{Q}_{11}^{nr}$ denote the maximal unramified extension of $\mathbb{Q}_{11}$. By the theory of tame ramification (see for example, [Se2]), we necessarily have $H\mathbb{Q}_{11}^{nr} = \mathbb{Q}_{11}^{nr}(q_1^{1/5})$, whence the assertion of Lemma 5.23 is now clear.

$\square$

We next discuss the unique prime of F above 5. Let w denote this unique prime and put $G_w = G(\overline{F}_w/F_w)$. We consider the natural map

$$g_5 \ : \ \mathrm{Hom}(G_w, \Omega) \to H^1(F_w, A_{1,5}).$$

Put $Y_5 = A_1(F_w)/5$, which we view, as usual, as a subgroup of $H^1(F_w, A_{1,5})$. Again if $\beta \in \mathrm{Hom}(G_w, \Omega)$, we write R_β for the fixed field of the kernel of β.

5.24. Lemma. *Let x be any element of the subgroup Y_5 of $H^1(F_w, A_{1,5})$. Let M_x denote the compositum of all fields R_β for β running over all elements of $\mathrm{Hom}(G_w, \Omega)$ with $g_5(\beta) = x$. Then M_x is an unramified extension of F_w.*

Proof. Let $\tilde{A}_1$ denote the reduction of A_1 modulo 5, and let $\tilde{A}_{1,5\infty}$ be the group of all 5-power torsion points on $\tilde{A}_1$. Let δ denote the canonical composite map

$$\delta \; : \; H^1(F_w, A_{1,5}) \to H^1(F_w, A_{1,5\infty}) \to H^1(F_w, \tilde{A}_{1,5\infty}).$$

The crucial point in the proof of the lemma is

$$\delta(Y_5) = 0. \tag{87}$$

Indeed, (87) is a special case of a very general principle (see [C-G] or [G2]). It can be justified by remarking that $\delta(Y_5)$ clearly maps to zero under the natural map from $H^1(F_w, \tilde{A}_{1,5\infty})$ to $H^1(F_w, \tilde{A}_1(\tilde{\mathbb{F}}_5))$. But the latter map is injective because $A_1(\tilde{\mathbb{F}}_5)$ is a torsion group. Let I_w denote the inertial subgroup of G_w. Now, analogous to (84), we have the commutative diagram

$$
\begin{array}{ccccc}
\mathrm{Hom}(G_w, \Omega) & \longrightarrow & H^1(F_w, A_{1,5}) & \overset{\delta}{\longrightarrow} & H^1(F_w, \tilde{A}_{1,5\infty}) \\
\downarrow & & & & \downarrow \\
\mathrm{Hom}(I_w, \Omega) & & \longrightarrow & & H^1(I_w, \tilde{A}_{1,5\infty}).
\end{array}
$$

Again the bottom horizontal arrow is injective, since Ω maps injectively under reduction modulo w and I_w acts trivially on Ω and $\tilde{A}_{1,5\infty}$. Hence, if β is any element of $\mathrm{Hom}(G_w, \Omega)$ with $g_5(\beta) \in Y_5$, it follows from (87) and the above commutative diagram that the restriction of β to I_w must be trivial, i.e. R_β is an unramified extension of F_w. This completes the proof of the lemma.

$$\square$$

5.25. Proof of Theorem 5.18. We can now complete the proof of Theorem 5.18. If $\beta \in \mathrm{Hom}(G_S, \Omega)$, we write K_β for the fixed field of $\mathrm{Ker}\ \beta$. Let r denote the dimension of $\mathcal{S}(A_1/F, 5)$ as a vector space over $\mathbb{F}_5$. We define Λ to be the compositum of all the fields K_β as β ranges over $g^{-1}(\mathcal{S}(A_1/F, 5))$. Since $\mathrm{Ker}\ g$ has order 5, we see immediately that $[\Lambda : F] = 5^{2r}$.

Now Lemma 5.22 shows that if we define P to be the compositum of $\mathbb{Q}(\mu_{11})^+$ and F,

$$P = \mathbb{Q}(\mu_{11})^+ F, \tag{88}$$

then we must have

$$g(\mathrm{Hom}(G(P/F), \Omega)) = \Omega \subset \mathcal{S}(A_1/F, 5).$$

This means that one of the fields K_β for β in $g^{-1}(\mathcal{S}(A_1/F, 5))$ is the field P. Hence $[\Lambda : P] = 5^{2r-1}$. But, combining Lemmas 5.23 and 5.24, we see immediately that Λ must be an unramified extension of P. Note that, since $r \geq 1$, this shows by class field theory that 5 must divide the class number of P. On the other hand, by (ii) of Proposition 5.26, 5 divides the class number of P to the first power only. Hence $r = 1$, and so $\mathcal{S}(A_1/P, 5) = \Omega$. In view of the exact sequence (80), this completes the proof of Theorem 5.18.

$\square$

Finally, we are left to prove the following proposition about the global arithmetic of F, which played a central role in the above descent argument.

5.26. Proposition. *(i) There is no non-trivial abelian 5-extension of* $F = \mathbb{Q}(\mu_5)$, *which is unramified outside the unique prime of F above 5, and in which every prime of F above 11 splits completely.*
(ii) The class number of the field $\mathbb{Q}(\mu_{55})$ is 10.

Proof. Assertion (ii) is very old, and is given, for example, in the table of class numbers at the end of [W]. Recall that Λ is the subfield of $\mathbb{Q}(\mu_{55})$ which is given by (88), namely it is the compositum of $\mathbb{Q}(\mu_5)$ and $\mathbb{Q}(\mu_{11})^+$. Since $[\mathbb{Q}(\mu_{55}) : \Lambda] = 2$, it follows from (ii) that the class number of Λ must be of the form $5 \cdot 2^k$ for some integer $k \geq 0$, and it is this latter fact which is needed in the proof of Theorem 5.18. We also remark in passing that the arguments in the proof of Theorem 5.18 show a priori that the class number of Λ, and therefore also the class number of $\mathbb{Q}(\mu_{55})$, must be divisible by 5. We do not see any elegant proof of (i), and we just blast it out by studying congruences in the Kummer generator. Thus we suppose that, contrary to (i), there is a cyclic extension L of F of degree 5, which is unramified outside the unique prime of F above 5, and in which the primes of F above 11 split completely. By Kummer theory, we then have $L = F(\beta^{1/5})$, where β

can only be divisible by the unique prime of F above 5. It follows from Hensel's lemma that a prime v of F above 11 will split completely in L if and only if the congruence $X^5 \equiv \beta \mod v$ is soluble in the ring of integers of F_v. This latter congruence will be soluble if and only if we have $\beta \equiv \pm 1 \mod v$. Let $\alpha_2 = (1 + \sqrt{5})/2$ be the fundamental unit of the maximal real subfield $\mathbb{Q}(\sqrt{5})$ of F. We define w to be one of the two primes of F above 11 such that $\alpha_2 \equiv 4 \mod w$. Write Δ for the Galois group $G(F/\mathbb{Q})$. Thus we conclude that the congruence

$$\sigma(\beta) \equiv \pm 1 \mod w \text{ for all } \sigma \in \Delta \tag{89}$$

must hold. Let ζ denote a primitive 5-th root of unity. Then the unique prime of F above 5 is generated by $\zeta - 1$. Thus, putting $\alpha_1 = \zeta$, $\alpha_2 = (1 + \sqrt{5})/2$, $\alpha_3 = \zeta - 1$, we see that we can write $\beta = \alpha_1^{n_1} \alpha_2^{n_2} \alpha_3^{n_3}$, where n_1, n_2, n_3 are integers, which we are only interested in modulo 5. We proceed to show by a brutal calculation that (89) yields four contradictory linear equations for n_1, n_2, n_3 modulo 5, unless all of n_1, n_2, n_3 are divisible by 5. Now 3 is of order 5 $\mod 11$, and so we fix our choice of ζ by assuming that $\zeta \equiv 3 \mod w$. Write σ_i $(i = 1, \dots, 4)$ for the element of Δ which acts on μ_5 by raising to the i-th power. The following table gives the values of the $\sigma(\alpha_j)$ modulo w:

	σ_1	σ_2	σ_3	σ_4
α_1	3	-2	5	4
α_2	4	-3	-3	4
α_3	2	3	4	3.

Taking the squares of the congruence (89), and noting that $4 \equiv 3^4 \mod 11$, and $5^2 \equiv 3 \mod 11$, we deduce immediately that the following four linear equations hold for n_1, n_2, n_3 viewed modulo 5:

$$2n_1 + 3n_2 + 4n_3 \equiv 0$$
$$4n_1 + 2n_2 + 2n_3 \equiv 0$$
$$n_1 + 2n_2 + 3n_3 \equiv 0$$
$$4n_1 + 4n_2 + n_3 \equiv 0.$$

But the only solutions of these equations are $n_1 \equiv n_2 \equiv n_3 \equiv 0 \mod 5$. This contradicts the hypothesis that L is an extension of F of degree 5, and so the proof of Proposition 5.26 is now complete.

$\square$

THE SELMER GROUP OF A_0 AND A_2 OVER $\mathbb{Q}(\mu_{5^\infty})$

5.27. Let $f_i(T)$ denote the characteristic power series for the dual of $\mathcal{S}(A_i/\mathbb{Q}(\mu_{5^\infty}))$. Theorem 5.4 shows that we can take $f_1(T) = 1$.

5.28. Theorem. *We have* $f_0(T) = 5^2$ *and* $f_2(T) = 5^4$.

Proof. We apply the formula of Perrin-Riou [P-R2] and Schneider [Sch] on the change of the characteristic power series under isogeny to the two isogenies

$$0 \longrightarrow \mu_5 \longrightarrow A_0 \longrightarrow A_1 \longrightarrow 0$$

$$0 \longrightarrow \mathbb{Z}/5 \longrightarrow A_0 \longrightarrow A_2 \longrightarrow 0.$$

Since μ_5 lies on the formal group of A_0 at the unique prime of F above 5, we conclude from the first exact sequence and the fact that $f_1(T) = 1$, that we can take $f_0(T) = 5^2$. Similarly, since $\mathbb{Z}/5$ does not lie on the formal group of A_0 at the unique prime above 5, we conclude from the second exact sequence and the fact that $f_0(T) = 5^2$, that we have $f_2(T) = 5^4$.

$\square$

5.29. Corollary. *We have*

$$\text{III}(A_0/F)(5) = 0, \ \# \ \text{III}(A_2/F)(5) = 5^4.$$

Proof. This follows immediately by combining Theorem 5.28 with the Euler characteristic formula (Theorem 3.3), recalling that (see Appendix, Proposition A.1.7)

$$f_i(0) = \chi(\Gamma, \mathcal{S}(A_i/K_\infty)) \ (0 \leq i \leq 2).$$

$\square$

THE CURVES OF CONDUCTOR 294 OVER $\mathbb{Q}(\mu_{7^\infty})$

5.30. Finally, we outline the arguments for studying the isogeny class of curves B_1 and B_2 (see 4.3) of conductor 294 over $\mathbb{Q}(\mu_{7^\infty})$. We now put

$$F = \mathbb{Q}(\mu_7), \ K_\infty = \mathbb{Q}(\mu_{7^\infty}),$$

and we recall that B_1 and B_2 have good ordinary reduction of F above 7. We shall assume without proof the following result of T. Fisher [F], which is parallel to Theorem 5.18 for the curve $A_1/\mathbb{Q}(\mu_5)$.

5.31. Theorem. *We have* $\mathcal{S}(B_1/F) = 0$ *when* $F = \mathbb{Q}(\mu_7)$ *and* $p = 7$.

It is remarkable that B_1/F satisfies Theorem 5.31 and also has $c_v = 1$ for all primes v of F dividing 2 and 3. As B_1 has good ordinary reduction at the unique prime of F above 7, we can apply Theorem 3.3 to it, with $p = 7$. Recalling that the reduction (50) of B_1 at the unique prime of F above 7 has 7 points in $\mathbb{F}_7$, we conclude from Theorem 5.31 and Theorem 3.3 that

$$\chi(\Gamma, \mathcal{S}(B_1/K_\infty)) = 1.$$

In particular, this shows that the characteristic power series of the dual of $\mathcal{S}(B_1/K_\infty)$ is a unit in $\Lambda(\Gamma)$, and so $\mathcal{S}(B_1/K_\infty)$ must be finite by the structure theory for finitely generated torsion modules over $\Lambda(\Gamma)$. We are very grateful to B. Totaro (unpublished) for providing us with the elegant proof given below of the following stronger result:

5.32. Theorem. *We have* $\mathcal{S}(B_1/K_\infty) = 0$.

Proof. We will use Theorem 3.11 to show that

$$H^1(\Gamma, \mathcal{S}(B_1/K_\infty)) = 0.$$

Since $\chi(\Gamma, \mathcal{S}(B_1/K_\infty)) = 1$, this will imply that $H^0(\Gamma, \mathcal{S}(B_1/K_\infty)) = 0$ and hence the assertion of the theorem. By Theorem 3.11, it suffices to show that the subgroup μ_7 of $B_1(F)(7)$ does not lie on the formal group of B_1 at the unique prime v of F above 7. If, on the contrary, μ_7 lay on the formal group, we would have an exact sequence of Galois modules

$$0 \longrightarrow \mu_7 \longrightarrow B_{1,7} \longrightarrow \tilde{E}(\mathbb{F}_7) \longrightarrow 0, \tag{90}$$

where $\tilde{E}$ denotes the curve (50) over $\mathbb{F}_7$. Now, as was explained in **4.3**, the action of $G(F/\mathbb{Q})$ on $\tilde{E}(\mathbb{F}_7)$ in (90) is non-trivial. Hence the existence of (90) would contradict the fact that, by the Weil pairing, the determinant of the Galois representation on $B_{1,7}$ is the character giving the action of $G(\overline{\mathbb{Q}}/\mathbb{Q})$ on μ_7. Hence μ_7 cannot lie on the formal group of B_1 at the unique prime of F above 7, proving Theorem 5.32.

$$\square$$

5.33. Theorem. *The characteristic power series of* $\mathcal{S}(B_2/\mathbb{Q}(\mu_{7\infty}))$ *is* 7^3.

Proof. We apply the formula of Perrin-Riou [P-R2] and Schneider [Sch] to the isogeny

$$0 \longrightarrow \mu_7 \longrightarrow B_1 \longrightarrow B_2 \longrightarrow 0$$

and use the fact that μ_7 does not lie on the formal group of B_1 at the unique prime of F above 7.

$\square$

5.34. Corollary. *We have* $\text{III}(B_2/F)(7) = 0$.

Proof. This follows immediately from Theorems 5.33 and 3.3.

$\square$

5.35. We make the following final remark. Theorem 5.4 shows that none of the curves A_0, A_1, A_2 of conductor 11 has a point of infinite order in the field $\mathbb{Q}(\mu_{5\infty})$. Similarly, Theorem 5.32 shows that none of the curves B_1, B_2 of conductor 294 has a point of infinite order in the field $\mathbb{Q}(\mu_{7\infty})$. However, it is unknown whether the first three curves have any point of infinite order in their common field of 5-power division points. Similarly, it is unknown whether either of the curves B_1, B_2 has a point of infinite order in their common field of 7-power division points. In view of Theorem 2.19, it seems very likely that points of infinite order exist in both situations, but it is a reflection on our lack of knowledge and lack of computing power to not be able to show this at present.

Appendix

In this appendix, we expand on some results that have been used in the notes. It is, of course, beyond the scope of these notes to provide detailed proofs of the results. However, we felt that it would be of use to the reader, if we isolated them and elaborated a little more, giving additional references to a fuller treatment.

A.1. STRUCTURE THEORY OF IWASAWA MODULES

A.1.1. The main reference for this section is Washington's book [W]. The results in this section have been used mainly in Chapters 2 and 3 of the notes.

Let F be a number field. We say a Galois extension F_∞/F is a $\mathbb{Z}_p$-extension if $\Gamma = G(F_\infty/F)$ is topologically isomorphic to $\mathbb{Z}_p$. Let $\Gamma_n = p^n\Gamma$ and write F_n for the fixed field of Γ_n, so that the degree of F_n over F is p^n.

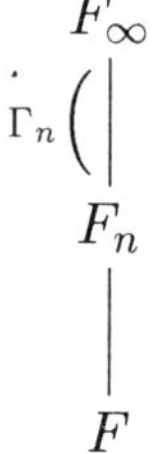

Recall that the Iwasawa algebra of Γ, denoted $\Lambda(\Gamma)$ (we shall abbreviate the notation to Λ) is defined as

$$\Lambda(\Gamma) := \varprojlim \mathbb{Z}_p[\Gamma/\Gamma_n],$$

where the inverse limit is defined via the ring homomorphism induced by the natural maps $G(F_{n+1}/F) \to G(F_n/F)$. There is an isomorphism

$$\mathbb{Z}_p[[T]] \simeq \Lambda$$

under which T maps to $(\gamma-1)$, where γ is any fixed topological generator of Γ. Thus Λ is a complete local noetherian ring.

Given a compact or discrete Γ-module M, the action of Γ can be extended to an action of the whole of Λ on M, by linearity and continuity. The structure of Λ-modules is well-understood, at least up to the weaker notion of pseudo-isomorphism. Recall that two Λ-modules M and M' are *pseudo-isomorphic* if there is a Λ-homomorphism $M \to M'$ with finite kernel and cokernel. If M and M' are pseudo-isomorphic, we write $M \sim M'$. The main structure theorem is the following, which was originally proved by Iwasawa, and later by Serre in a more direct fashion.

A.1.2. Theorem. *Let M be a finitely generated Λ-module. Then*

$$M \sim \Lambda^r \oplus \left(\bigoplus_{i=1}^{s} \Lambda/p^{n_i} \right) \oplus \left(\bigoplus_{j=1}^{t} (\Lambda/f_j(T)^{m_j}) \right)$$

where r, s, t, n_i, $m_j \in \mathbb{Z}$ and $f_j(T) \in \mathbb{Z}_p[T]$ are irreducible monic distinguished polynomials.

(A monic polynomial $f(T) \in \mathbb{Z}_p[T]$ is *distinguished* if p divides all its coefficients except the leading one).

The r that appears in such an expression is the *rank* of M as a Λ-module. The module M is Λ-torsion if and only if $r = 0$. Put

$$g(T) = \left(\prod_i p^{n_i} \right)\left(\prod_j f_j(T)^{m_j} \right).$$

If $r = 0$, we define the *characteristic power series* of M to be any generator of the principal ideal $(g(T))$ in Λ. Note that a characteristic power series is a well defined element of Λ, up to mulitiplication by a unit. Assuming $r = 0$, the integer $\sum_{i=1}^{s} n_i$ is called the μ-*invariant* of M and the integer $\sum_{j=1}^{t} m_j \deg(f_j(T))$ is called the λ-*invariant* of M. We now state an analogue of Nakayama's lemma for Λ-modules. We refer the reader to [W] for a proof, or to [B-H] for a detailed proof in a much more general context.

A.1.3. Lemma. *Let M be a compact Λ-module and $\mathfrak{m}$ the unique maximal ideal of Λ. If $\mathfrak{m}M = M$, then $M = 0$.*

A.1.4. Corollary. *Let M, $\Lambda, \mathfrak{m}$ be as above. If $M/\mathfrak{m}M$ is finitely generated as a $\Lambda/\mathfrak{m}(= \mathbb{Z}/p)$-module, then M is finitely generated as a Λ-module.*

Proof. The hypothesis implies that $M/\mathfrak{m}M$ is finite as a $\Lambda/\mathfrak{m}$-module. Let $m_1, \ldots, m_n \in M$ be elements of M which generate $M/\mathfrak{m}M$ and let $N = <m_1, \ldots, m_n>$. Then $N + \mathfrak{m}M = M$ and M/N is compact. But $\mathfrak{m}(M/N) = M/N$, and so Lemma A.1.3 implies that $M = N$ and hence M is finitely generated as a Λ-module.

$\square$

A.1.5. Lemma. *Let A be a discrete p-primary Γ-module and suppose $\widehat{A}$, its Pontryagin dual, is a finitely generated Λ-module of rank r. Then the following are equivalent:*

(i) $H^1(\Gamma, A) = 0$

(ii) $(\widehat{A})^{\Gamma} = 0$

(iii) $(\widehat{A})_{\Gamma}$ *has $\mathbb{Z}_p$-rank r and $\widehat{A}$ has no non-zero finite submodule.*

Proof. $(i) \Longleftrightarrow (ii)$: This follows from a standard duality result in group cohomology.

$(i) \Longleftrightarrow (iii)$: We first observe the following general reuslt which follows from the structure theory of Λ-modules and the identification of $\mathbb{Z}_p[[T]]$ with Λ under which $T \mapsto \gamma - 1$, γ a topological generator of Γ. Suppose M is a finitely generated Λ-module such that

$$M \sim \Lambda^k \oplus (\oplus_i (\Lambda/(f_i(T))))$$

where $f_i(T) \in \mathbb{Z}_p[T]$ is a monic distinguished polynomial or a constant of the form p^m. Clearly, $(\Lambda)_{\Gamma} \simeq \mathbb{Z}_p$ and

$$(\Lambda/f_i(T))_{\Gamma} \simeq \mathbb{Z}_p \Longleftrightarrow T \text{ divides } f_i(T).$$

Thus,

$$\mathbb{Z}_p - \text{rank of } M_{\Gamma} = k + \delta$$

where

$$\delta = \#\{i \mid f_i(0) = 0\}. \tag{A1}$$

Now, since T maps to $\gamma - 1$, if $x \in M^{\Gamma}$, then $\gamma(x) = x$ and hence $(\gamma - 1)x = Tx = 0$. It is therefore clear that

$$Z \simeq \Lambda/(T) \Longrightarrow Z^{\Gamma} \simeq \mathbb{Z}_p,$$

and

$$Z \simeq (\Lambda/p^m) \Longrightarrow Z^{\Gamma} = 0.$$

If Z is of the form $\Lambda/f_i(T)$ where $f_i(T)$ is a monic distinguished polynomial, then one sees that

$$\mathbb{Z}_p[T]/(f_i(T)) \simeq \mathbb{Z}_p[[T]]/(f_i(T))$$

and therefore $0 \neq x \in Z^\Gamma \Longleftrightarrow T \mid f_i(T)$. Noting that pseudo-isomorphism does not affect the $\mathbb{Z}_p$-rank, it follows that

$$\mathbb{Z}_p - \text{rank of } M^\Gamma = \delta$$

where δ is as in (A1) above. The implications $(i) \Longleftrightarrow (iii)$ is now an easy exercise, on taking $M = \hat{A}$.

$\square$

The arguments in the above proof give the following result.

A.1.6. Corollary. *Let M, Λ, Γ be as before. Then*

$$M_\Gamma \, finite \Longrightarrow M \text{ is } \Lambda - \text{torsion}.$$

$\square$

We mention below how the characteristic power series of a torsion $\Lambda(\Gamma)$-module relates to the Γ-Euler characteristic of its Pontryagin dual. Let M be a finitely generated Λ-torsion module, and let $f_M(T)$ be a characteristic power series of M. Then $\hat{M} = \text{Hom}(M, \mathbb{Q}_p/\mathbb{Z}_p)$ is a discrete Γ-module.

A.1.7. Proposition. *With notation as above, the following assertions hold:*

(i) $f_M(0) \neq 0$ *if and only if either $H^0(\Gamma, \hat{M})$ or $H^1(\Gamma, \hat{M})$ is finite.*

(ii) *If $f_M(0) \neq 0$, we have*

$$\mid f_M(0) \mid_p^{-1} = \# \, H^0(\Gamma, \hat{M})/\# \, H^1(\Gamma, \hat{M}).$$

The proof is an easy consequence of the structure theory of Λ-modules (see [G2, §4] for a detailed proof). In the notes, we use this proposition for $M = \mathcal{S}(\widehat{E/K_\infty})$, where E is an elliptic curve over a number field F having good ordinary reduction at all primes of F dividing p, and such that the selmer group $\mathcal{S}(E/F)$ is finite, with K_∞ the cyclotomic $\mathbb{Z}_p$-extension of F.

A.2. Deeply ramified p-adic fields and cyclotomic extensions

A.2.1. One of the chief inputs in the local cohomology calculations in Chapter 3 is the theory of deeply ramified extensions. The main reference for this is [C-G]. We shall state the principal results from this theory that are used in these notes.

Let $\overline{\mathfrak{M}}$ denote the maximal ideal of the ring of integers of $\overline{F_v}$. Suppose K/F_v is an algebraic extension. Then the Galois group $G(\overline{F_v}/K)$ acts on $\overline{\mathfrak{M}}$, and we have the cohomology groups $H^i(K, \overline{\mathfrak{M}}) = H^i(G(\overline{F_v}/K), \overline{\mathfrak{M}})$.

A.2.2. Definition. An algebraic extension K of $\mathbb{Q}_p$ is said to be *deeply ramified* if $H^1(K, \overline{\mathfrak{M}}) = 0$.

There are various equivalent conditions that define a deeply ramified extension (cf. [C-G, p. 143]), but for the purposes of these notes, the cohomological condition above is the most useful.

Let L be any finite extension of $\mathbb{Q}_p$. A basic example of a deeply ramified field K is a ramified $\mathbb{Z}_p$-extension of L. More generally, a large class of examples occuring from a theorem of Sen are Galois extensions K of L such that $G(K/L)$ is a p-adic Lie group with infinite inertial subgroup [C-G, Theorem 2.13]. Let E/L be an elliptic curve. Recall that associated to E are two natural objects, namely $\hat{E}_v$ which is the formal group over the ring of integers of L and $\tilde{E}_v$ which is the reduced curve, defined over the residue field k_v (cf. [Si, Chapters IV and VII]). The formal group is in fact the kernel of the reduction modulo v. The principal result of [C-G] which is used in these notes is the following (cf. [C-G, Theorem 3.1 and Corollary 3.2]).

A.2.3 Theorem. *Let E/L be an elliptic curve. for each deeply ramified extension K of L, we have*

$$H^i(K, \hat{E}_v(\overline{\mathfrak{M}})) = 0 \ \forall \ i \geq 1.$$

A.2.4. One of the main advantages of working with deeply ramified extensions is that it vastly simplifies the computations of local cohomology groups whose coefficient modules are not torsion groups. This is amply borne out in the local computations of Chapter 3 for a prime v dividing p such that E has good ordinary reduction at v. The deeply ramified extension of F_v that one works with here is $K_{\infty,v}$, the cyclotomic $\mathbb{Z}_p$-extension of F_v. Suppose E/F_v and $K_{\infty,v}$ are as above. Then we have (cf. [C-G, Proposition 4.8])

$$H^1(K_{\infty,v}, E(\overline{F_v}))(p) \simeq H^1(K_{\infty,v}, \tilde{E}_{p^\infty}). \qquad (A2)$$

The main ingredient in proving (A2) is the explicit description of the image of the Kummer homomorphism for deeply ramified extensions (cf. [C-G, Proposition 4.3]).

A.2.5. A second striking application of the theory of deeply ramified extensions is in the study of universal norm subgroups of elliptic curves [C-G, §5]. As before, F_v is a finite extension of $\mathbb{Q}_p$ and E/F_v an elliptic curve. For any profinite abelian group X, let X^* denote its maximal pro-p subgroup.

Suppose K/F_v is an algebraic extension. Recall that the group of universal norms of E from K is defined as the subgroup

$$E_U(K) \;=\; N_{K/F_v}(E) = \bigcap_L N_{L/F_v}(E(L))$$

where L runs over all finite extensions of F_v in K, and N_L/F_v denotes the norm map. We have the non-degenerate Tate pairing

$$H^1(F_v, E)(p) \times E^*(F_v) \to \mathbb{Q}_p/\mathbb{Z}_p,$$

and the restriction map

$$r : H^1(F_v, E)(p) \to H^1(K, E)(p).$$

It is very well-known that $E_U^*(K)$ is the exact orthogonal complement of Ker r in the Tate pairing. However, it is not a priori easy to determine either of these groups without the help of the theory of deeply ramified fields. We illustrate what happens when we assume that E has good ordinary reduction over F_v and K is any deeply ramified extension of F_v. Indeed, under the above hyotheses, in the notation of [C-G], we then have

$$C = \hat{E}_{p^\infty}, \; D = \tilde{E}_{p^\infty},$$

where $\hat{E}$ (resp. $\tilde{E}$) is the associated formal group (resp. reduced curve). Further, the homomorphism

$$\pi_{F_v} \; : \; H^1(F_v, E_{p^\infty}) \; \to \; H^1(F_v, \tilde{E}_{p^\infty})$$

is surjective (cf. [C-G], Proposition 5.3]). Let

$$\rho \; : \; H^1(F, \tilde{E}_{p^\infty}) \to H^1(K, \tilde{E}_{p^\infty})$$

denote the restriction map. As K/F_v is a deeply ramified extension, we see from [C-G, Theorem 5.2] that the group

$$\Omega(E, K/F_v) := \; H^1(F_v, \tilde{E}_{p^\infty})/\operatorname{Ker} \, \rho$$

is canonically isomorphic to the image of the restriction map

$$r : H^1(F_v, E)(p) \; \to \; H^1(K, E)(p).$$

From this, one deduces that the subgroup $E_U^*(K)$ of $E(F_v)$ is canonically dual to $\Omega(E, K/F_v)$. This gives the desired description of $E_U^*(K)$ in terms of the cohomology of the p-primary Galois module $\tilde{E}_{p^\infty}$.

A.2.6. In this paragraph, we discuss briefly the cyclotomic $\mathbb{Z}_p$-extension of a number field F. The main reference for this is [W], see also [L]. Let $F(\mu_{p^\infty})$ be the Galois extension of F obtained by adjoining the p-power roots of unity. Recall that the cyclotomic $\mathbb{Z}_p$-extension of F, denoted K_∞, is by definition, the fixed field of the torsion subgroup of $G(F(\mu_{p^\infty})/F)$. The extension K_∞/F is unramified at all primes v of F which do not lie above p. On the other hand, every prime lying above p ramifies in K_∞. Further, there exists $n \geq 0$ such that every prime which ramifies in K_∞/K_n is totally ramified. We note there are only finitely many primes that lie above p. More generally, it is also easy to see that there are only finitely many primes of K_∞ above every rational prime. If w is any prime of K_∞, we write $K_{\infty,w}$ for the union of the completions of all the finite extensions of F contained in K_∞. It is plain now that if w is a prime of K_∞ that does not lie above p, then the extension $K_{\infty,w}$ is the unique unramified $\mathbb{Z}_p$-extension of the completion of F at w. If w divides p, then $K_{\infty,w}$ is one of the basic examples of a deeply ramified field.

A.2.7. In this paragraph, we discuss Imai's theorem and give a non-standard proof for elliptic curves using Lie algebra cohomology. Our main motivation for this is that it leads to a simple proof of Serre's theorem (see Theorem 3.19) on the Euler characteristics of elliptic curves. Let us also recall that we used Imai's theorem in Chapter 3, to prove the finiteness of Ker β in the fundamental diagram.

Let A be an abelian variety defined over a finite extension L of $\mathbb{Q}_p$, and let L_∞ be the cyclotomic $\mathbb{Z}_p$-extension of L. Then, Imai proved that the p-primary subgroup of $A(L_\infty)$ is finite, provided that A has potential good reduction. We will only need the following global statement for an elliptic curve E defined over a finite extension F of $\mathbb{Q}$.

A.2.8. Theorem. *The group $E_{p^\infty}(K_\infty)$ is finite, where K_∞ is the cyclotomic $\mathbb{Z}_p$-extension of F.*

Proof. Consider the tower of extensions

$$
\begin{array}{c}
F_\infty \\
\Delta\left(\;\Big|\;\right. \\
K_\infty \\
\Gamma\left(\;\Big|\;\right. \\
F
\end{array}
$$

where we recall that $F_\infty = F(E_{p^\infty})$. Let $\Sigma = G(F_\infty/F)$ and $\Delta = G(F_\infty/K_\infty)$. As usual, we define

$$
T_p(E) = \varprojlim_{n} E_{p^n},
$$

and

$$
V_p(E) = T_p E \otimes \mathbb{Q}_p.
$$

The module $T_p E$ is a free $\mathbb{Z}_p$-module of rank two and the action of Σ on it gives an injective homomorphism

$$
\Sigma \hookrightarrow GL_2(\mathbb{Z}_p)
$$

via which we can identify Σ as a compact subgroup of $GL_2(\mathbb{Q}_p)$. Hence Σ is a p-adic Lie group. To prove the assertion of the theorem, we can clearly assume that $\mu_p \subset F$ if $p > 2$ and $\mu_4 \subset F$ if $p = 2$, so that $K_\infty = F(\mu_{p^\infty})$. By the Weil pairing, $\Delta = G(F_\infty/K_\infty)$ is given by the intersection of $SL_2(\mathbb{Z}_p)$ with Σ. Let $\mathfrak{d}$ denote the Lie algebra of Δ. We recall that $H^0(\mathfrak{d}, V_p(E))$ is the $\mathbb{Q}_p$-subspace of all vectors in $V_p(E)$ which are annihilated by all elements in the Lie algebra $\mathfrak{d}$.

The starting point of our proof is to show that

$$
H^0(\mathfrak{d}, V_p(E)) = 0. \tag{A3}
$$

To do this, it clearly suffices to prove that there exist invertible elements in $\mathfrak{d}$. If E does not have complex multiplication, we appeal to Serre's deep theorem that Σ is open in $GL_2(\mathbb{Z}_p)$, whence Δ is open in $SL_2(\mathbb{Z}_p)$. Thus, in this case, $\mathfrak{d}$ is the Lie algebra of $SL_2(\mathbb{Z}_p)$ which is well-known to consist of all matrices in $M_2(\mathbb{Q}_p)$ with trace zero. Therefore $\mathfrak{d}$ clearly has invertible elements. Assume now that E has complex multiplication. We use the local results given in the appendix of Serre [Se6] to show the existence of an invertible element in $\mathfrak{d}$. Pick any place v of F lying above p. Since E automatically has potential good reduction at v (because E has complex multiplication), the Lie algebra of the decomposition group of v is given by Theorem A.2.2 of [Se6] when E has potential supersingular reduction at v, and by Corollary 2 of A.2.4. in [Se6] when E has potential ordinary reduction at v. It is clear from the explicit description of these Lie algebras that they always contain invertible elements of trace zero. But the Lie algebra of the decomposition group is a subalgebra of $\mathfrak{d}$. This completes the proof of (A3).

We now appeal to results from the cohomology of Lie algebras to conclude from (A3) that

$$H^i(\mathfrak{d}, V_p(E)) = 0 \ \forall \ i \geq 0. \tag{A4}$$

By a well-known result in the cohomology of Lie algebras [H-S, Theorem 10], it suffices to show that $\mathfrak{d}$ is a reductive Lie algebra and that $V_p(E)$ is a semisimple $\mathfrak{d}$-module. This is clear when E does not have complex multiplication, as $\mathfrak{d}$ is then equal to the Lie algebra of $SL_2(\mathbb{Z}_p)$ which is well-known to be semisimple. When E has complex multiplication, $V_p(E)$ is a faithful semisimple representation of the group Σ and hence also for the Lie algebra $\mathfrak{g}$ of Σ. This shows that $\mathfrak{g}$ is indeed a reductive Lie algebra, whence $\mathfrak{d}$ is also, because it is an ideal in $\mathfrak{g}$. Also, $V_p(E)$ is a semisimple $\mathfrak{d}$-module. Hence (A4) follows from (A3).

By a deep theorem of Lazard [La, V.2.4.10], the group $H^i(\Delta, V_p(E))$ is a $\mathbb{Q}_p$-subspace of $H^i(\mathfrak{d}, V_p(E))$ for all $i \geq 0$. Hence

$$H^i(\Delta, V_p(E)) = 0 \ \forall \ i \geq 0. \tag{A5}$$

Consider the short exact sequence

$$0 \longrightarrow T_p(E) \longrightarrow V_p(E) \longrightarrow E_{p^\infty} \longrightarrow 0$$

of Galois modules. Now since $H^i(\Delta, V_p(E)) = 0$, we see that the group $H^i(\Delta, T_p(E))$ is finite for all i. Indeed $H^i(\Delta, T_p(E))$ is a $\mathbb{Z}_p$-module of finite type [La, V.3.2.7], isomorphic to $H^{i-1}(\Delta, E_{p^\infty})$. But this latter group is torsion and therefore $H^i(\Delta, T_p(E))$ is necessarily finite for all i, and

$$H^i(\Delta, E_{p^\infty}) \text{ is finite } \forall \, i \geq 0. \tag{A6}$$

The case $i = 0$ is Theorem A.2.8. The finiteness of the higher cohomology groups does not seem to have been pointed out before (see [C-S]).

$$\square$$

A.2.9. In fact, the above argument leads to a new and simple proof of Theorem 3.19, which has the merit of generalizing almost immediately to all abelian varieties over a number field (see [C-S]). Again, let K_∞ denote the cyclotomic $\mathbb{Z}_p$-extension of F and $\Gamma = G(K_\infty/F)$. Then Γ has p-cohomological dimension equal to 1, and the Hochschild-Serre spectral sequence yields the short exact sequence, for all $i \geq 1$,

$$0 \longrightarrow H^1(\Gamma, H^{i-1}(\Delta, E_{p^\infty})) \longrightarrow H^i(\Sigma, E_{p^\infty})$$
$$.0 \longleftarrow H^0(\Gamma, H^i(\Delta, E_{p^\infty}))$$

It was shown in (A6) that $H^i(\Delta, E_{p^\infty})$ is finite for all $i \geq 0$ and all primes p. Moreover, if Y is any finite group on which Γ acts, the groups $H^0(\Gamma, Y)$ and $H^1(\Gamma, Y)$ have the same cardinality. Hence, if we let h_i denote the order of $H^0(\Gamma, H^i(\Delta, E_{p^\infty}))$, we conclude from the above exact sequence that

$$\#(H^i(\Sigma, E_{p^\infty})) = h_i h_{i-1} \, \forall \, i \geq 1. \tag{A7}$$

Now assume that p is such that Σ has no element of order p. By results of Lazard and Serre (cf. [Se3]), the p-cohomological dimension of Σ is finite and equal to the dimension d of Σ as a p-adic Lie group. We can then define

$$\chi(\Sigma, E_{p^\infty}) = \prod_{i=0}^{d} (\# \, H^i(\Sigma, E_{p^\infty}))^{(-1)^i}.$$

It is clear from (A7) that

$$\chi(\Sigma, E_{p^\infty}) = h_0 \cdot (h_0 h_1)^{-1} \cdot (h_1 h_2) \cdots = 1.$$

To complete the proof of Theorem 3.19, it remains to show that

$$H^d(\Sigma, E_{p^\infty}) = 0.$$

Choose n so large that p^n annihilates $H^d(\Sigma, E_{p^\infty})$. Taking cohomology of the exact sequence

$$0 \longrightarrow E_{p^n} \longrightarrow E_{p^\infty} \xrightarrow{\;p^n\;} E_{p^\infty} \longrightarrow 0$$

and recalling that $H^{d+1}(\Sigma, E_{p^\infty})$ is zero, it follows immediately that $H^d(\Sigma, E_{p^\infty}) = 0$, as required.

A.3. TATE PARAMETRIZATION OF ELLIPTIC CURVES

The main references for this are [R] and [Ta2]. Suppose E is an elliptic curve over $\mathbb{C}$. It is well-known that there is an analytic isomorphism between the group $E(\mathbb{C})$ and $\mathbb{C}/\Lambda$, where Λ is a lattice in $\mathbb{C}$ (i.e., a discrete subgroup of $\mathbb{C}$ which contains an $\mathbb{R}$-basis). Let now E be a curve over a finite extension F_v of $\mathbb{Q}_p$. When considering p-adic elliptic curves, there is no such parametrization of $E(F_v)$ as a quotient of the additive group of F_v. On the other hand there is the normalized exponential map

$$\exp : \mathbb{C}/\Lambda \to \mathbb{C}^*/q^{\mathbb{Z}}$$

where if $\Lambda = \mathbb{Z} + \tau\mathbb{Z}$, $q = \exp(\tau)$. Now discrete subgroups of F_v^* abound. Tate showed that if $q \in F_v^*$ is such that $0 \neq |q| < 1$, then $F_v^*/q^{\mathbb{Z}} =: E_q$ is an elliptic curve over F_v. More precisely, Tate showed the following: Let q be any non-zero element of F_v with $\mathrm{ord}_v(q) > 0$. Define

$$a_4 = -5 \sum_{n \geq 1} n^3 q^n/(1 - q^n), \quad a_6 = 1/12 \sum_{n \geq 1} (7n^5 + 5n^3)q^n/(1 - q^n).$$

Then

$$E_q \; : \; y^2 + xy = x^3 + a_4 x + a_6$$

is an elliptic curve over F_v which we call the *Tate curve* attached to q. We call q the *Tate period* of E_q. It can be shown that $\mathrm{ord}_v(j(E_q)) < 0$ and that E_q has split multiplicative reduction.

Theorem A.3.1. *Let q be any element of F_v^* with $\mathrm{ord}_v(q) > 0$. Then there is an analytic isomorphism*

$$\phi \; : \; F_v^*/q^{\mathbb{Z}} \; \simeq \; E_q(F_v).$$

Conversely, if E is any elliptic curve over F_v with $\mathrm{ord}_v(j_E) < 0$ and split multiplicative reduction, then E is isomorphic over F_v to a Tate curve E_q for some q in F_v with $\mathrm{ord}_v(q) > 0$.

This parametrization was used in Chapter 5 in the explicit calculations of universal norms for the curves of conductor 11 and 294.

A.4. MULTIPLICATIVE KUMMER GENERATORS

In this section, we record some well-known results on the Weil pairing and multiplicative Kummer theory. All this is folklore (Weil, Cassels, Tate, ...), and the results below have been used in the cup-product computation of Chapter 5. Let F be any field of characteristic zero, and E/F an elliptic curve. Let m be an integer > 1, and E_m the group of m-division points in $E(\overline{F})$. We assume that $E_m \subset E(F)$.

A.4.1. Lemma. *There is a canonical isomorphism*

$$\phi \; : \; H^1(F, E_m) \;\simeq\; \mathrm{Hom}(E_m, F^*/F^{*^m}).$$

Proof. Put $G_F = G(\overline{F}/F)$. For simplicity, we shall write '=' to denote a canonical, obvious isomorphism. Since $E_m \subset E(F)$, we have $H^1(F, E_m) = \mathrm{Hom}(G_F, E_m)$. But the Weil pairing gives a canonical isomorphism $E_m = \mathrm{Hom}(E_m, \mu_m)$. Hence

$$H^1(G_F, E_m) = \mathrm{Hom}(G_F, \mathrm{Hom}(E_m, \mu_m)) = \mathrm{Hom}(E_m, \mathrm{Hom}(G_F, \mu_m)).$$
$$(A8)$$

But $\mu_m \subset F$ because $E_m \subset E(F)$, and so, multiplicative Kummer theory gives the identification $F^*/F^{*^m} = \mathrm{Hom}(G_F, \mu_m)$, and the proof is complete.

$\square$

The next corollary spells out how we get the multiplicative Kummer generators corresponding to a given $\alpha \in H^1(F, E_m)$.

A.4.2. Corollary. *Assume that $E_m \subset E(F)$. Let α be any element of $H^1(F, E_m) = \mathrm{Hom}(G_F, E_m)$, and let L_α be the fixed field of $\mathrm{Ker}\ \alpha$. Then*

$$L_\alpha = F(\phi(\alpha)(u)^{1/m},\ u \in E_m). \qquad (A9)$$

Proof. Since α factors through L_α, it is clear that the image of α under the chain of canonical isomorphisms in (A8) lands in the group $\mathrm{Hom}(E_m, \mathrm{Hom}(G(L_\alpha/F), E_m))$. Hence $\phi(\alpha)(u)$ must be contained in the Kummer group of L_α for all $u \in E_m$. Thus the right hand side of (A9) is contained in the left hand side. Conversely, it is clear that α must factor through the field on the right hand side of (A9), and so this field must contain L_α.

$\square$

A.4.3. Finally, we briefly state the computation of the cup-product pairing in terms of Hilbert norm residue symbols. Assume now that F is an extension of $\mathbb{Q}_p$ for some prime p. We continue to assume $E_m \subset E(F)$, whence $\mu_m \subset F$. Now we have three different pairings, namely:

(i) Weil pairing $\quad w_m \; : \; E_m \times E_m \; \to \; \mu_m$

(ii) Norm residue symbol $\quad (\, , \,)_m \; : \; F^*/F^{*m} \times F^*/F^{*m} \; \to \; \mu_m$

(iii) Tate pairing $\quad (\, , \,) \; : \; H^1(F, E_m) \times H^1(F, E_m) \; \to \; \mathbb{Z}/m\mathbb{Z}.$

The precise relation between the three pairings is the following.

A.4.4. Lemma. *For all u_1, u_2 in E_m, and all α_1, α_2 in $H^1(F, E_m)$, we have*

$$w_m(u_1, u_2)^{(\alpha_1, \alpha_2)} \;=\; (\phi(\alpha_1)(u_1), \phi(\alpha_2)(u_2))_m.$$

$\square$

Bibliography

[B-H] Balister, P., Howson, S., *Note on Nakayama's lemma for compact Λ-modules*, Asian J. of Math. **1** (1997), 214–219.

[B-K] Bloch, S., Kato, K., *L-functions and Tamagawa numbers of motives*, in *The Grothendieck Festschrift vol 1*, (1990), Progress in Math. 86, Birkhaüser, 333–400.

[Ca] Cassels, J., *The dual exact sequence*, Crelle **216** (1964), 150–158.

[C] Coates, J., *Fragments of the GL_2 Iwasawa theory of elliptic curves without complex multiplication*, Cetraro notes (To appear).

[C-G] Coates, J., Greenberg, R., *Kummer theory for abelian varieties over local fields*, Invent. Math. **124** (1996), 129–174.

[C-H] Coates, J., Howson, S., *Euler characteristics of elliptic curves II*, (To Appear).

[C-M] Coates, J., McConnell, G., *Iwasawa theory of modular elliptic curves of analytic rank atmost* 1, Jour. London Math. Soc. **(2) 50** (1994), 243–264.

[C-S] Coates, J., Sujatha, R., *Euler-Poincaré characteristics of abelian varieties*, To appear in C.R.A.S. Paris.

[Cr] Cremona, J., *Algorithms for modular elliptic curves*, Second Edition, Cambridge University Press, 1997.

[F] Fisher, T., *Ph.D Thesis*, Cambridge University.

[G] Greenberg R., *Iwasawa theory of p-adic representations*, Adv. Studies in Pure Math. **17** (1989), 97–137.

[G1] Greenberg, R., *The structure of Selmer groups* Proc. Nat. Acad. Sc. **94** (1997), 11125–11128.

[G2]   Greenberg, R., *Iwasawa theory for elliptic curves*, Cetraro notes (To appear).

[Gr]   Gross, B., *Kolyvagin's work on modular elliptic curves*, in *L*-functions and arithmetic, London Math. Soc. Lecture Note Ser. **153** (1991), Cambridge Univ. Press, Cambridge, 235–256.

[Ha]   Haberland, K., *Galois cohomology of algebraic number fields*, VEB Deutscher Verlag der Wissenschaften, Berlin, 1978.

[H1]   Harris, M., *p-adic representations arising from descent on abelian varieties*, Comp. Math. **39** (1979), 177–245.

[H]   Howson, S., *Iwasawa theory of elliptic curves for p-adic Lie-extensions*, Ph.D. Thesis, Cambridge University (1998).

[H-M]   Hachimori, Y., Matsuno, K., *An analogue of Kida's formula for the Selmer group of elliptic curves*, Jour. Alg. Geom. **8** (1999), 581–600.

[H-S]   Hochschild, G., Serre, J.-P., *Cohomology of Lie algebras*, Ann. of Math. **57** (1953), 591–603.

[I]   Imai, H., *A remark on the rational points of abelian varieties with values in cyclotomic $\mathbb{Z}_l$-extensions*, Proc. Japan. Acad. Math. Sci. **51** (1975), 12–16.

[K]   Kato, K., *p-adic Hodge theory and values of zeta functions of modular forms* (To appear).

[Ko]   Kolyvagin, V., *Euler Systems*, in *The Grothendieck Festscrift vol 2* (1990), Progress in Math., Birkhaüser, 435–483.

[L]   Lang, S., *Cyclotomic fields I and II*, Graduate Texts in Mathematics, Springer Verlag, New York, 1990.

[La]   Lazard, M., *Groupes analytiques p-adiques* Publ. Math. IHES **26** (1965), 389–603.

[Ma]   Mazur, B., *Rational points of abelian varieties with values in towers of number fields*, Invent. Math. **18** (1972), 183–266.

[Mc]   McCallum, W., *Tate duality and wild ramification*, Math. Ann. **288** (1990), 553–558.

[Mi] Milne, J., *Arithmetic Duality theorems*, Perspectives in Math. Vol 1, Academic Press, 1986.

[P-R] Perrin-Riou, B., *Théorie d'Iwasawa et hauteurs p-adiques*, Invent. Math. **109** (1992), 137–185.

[P-R2] Perrin-Riou, B., *Fonctions L p-adiques, Théorie d'Iwasawa et Points de Heegner*, Bull. Soc. Math. France **115 (4)** (1987), 399–456.

[P] Poitou, G., *Cohomologie Galoisienne des Modules Finis*, Dunod, Paris, 1967.

[R] Robert, A., *Elliptic curves*, Lecture Notes in Math. Vol 326, Springer, 1973.

[Sch] Schneider, P., *The μ-invariant of isogenies*, Jour. Indian Math. Soc. **52** (1987), 159–170.

[Se] Serre, J.-P., *Cohomologie Galoisienne, Cinquième édition*, Lecture notes in Math. Vol 5, Springer, 1994.

[Se1] Serre, J.-P., *Classes des Corps Cyclotomiques (d'après K. Iwasawa)*, Séminaire Bourbaki Exposé **174** (1958-59).

[Se2] Serre, J.-P., *Propriétés Galoisienne des points d'ordre fini des Courbes Elliptiques*, Invent. Math. **15** (1972), 259–331.

[Se3] Serre, J.-P., *Sur la dimension cohomologique des groupes profinis*, Topology **3** (1965), 413–420.

[Se4] Serre, J.-P., *Sur les groupes de congruence des variétés abéliennes II*, Izv. Akad. Nauk. SSSR. **35** (1971), 731–735.

[Se5] Serre, J.-P., *La distribution d'Euler Poincaré d'un groupe profini*, *in* Galois representations in arithmetic algebraic geometry **254** (1998), London Math. Soc. Lecture Note Ser., 461–493.

[Se6] Serre, J.-P., *Abelian l-adic representations and elliptic curves*, Benjamin 1968.

[Si] Silverman, J., *The Arithmetic of Elliptic curves*, Graduate Texts in Maths 106, Springer, 1986.

[Si2]   Silverman, J., *Advanced topics in the arithmetic of elliptic curves*, Graduate Texts in Maths 151, Springer, 1994.

[Ta]   Tate, J., *WC groups over p-adic fields*, Séminaire Bourbaki, Exposé 156 (1957-58).

[Ta2]   Tate, J., *A review of non-archimedean elliptic functions*, in Elliptic curves, Modular forms and Fermat's last theorem 2nd ed. (1997), International Press, 310–314.

[W]   Washington, L., *Introduction to Cyclotomic fields*, Graduate Texts in Maths 83, Springer, 1986.

[Wi]   Wiles, A., *Modular elliptic curves and Fermat's Last Theorem*, Ann. of Math. **141** (1995), 443–551.

Tata Institute of Fundamental Research
Lectures on Mathematics
Editor: S. Ramanan

	Title	Author
1.	On the Riemann Zeta-Function*	K. Chandrasekharan
2.	On Analytic Number Theory*	H. Rademacher
3.	On Siegel's Modular Functions*	H. Maass
4.	On Complex Analytic Manifolds	L. Schwartz
5.	On the Algebraic theory of Fields*	K.G. Ramanathan
6.	On Sheaf Theory*	C.H. Dowker
7.	On Quadratic Forms	C.L. Siegel
8.	On Semi-group Theory and its Application to Cauchy's Problem in Partial Differential Equations*	K. Yosida
9.	On Modular Correspondences*	M. Eichler
10.	On Elliptic Partial Differential Equations*	J.L. Lions
11.	On Mixed Problems in Partial Differential Equations and Representations of Semi-groups*	L. Schwartz
12.	On Measure Theory and Probability*	H.R. Pitt
13.	On the Theory of Functions of Several Complex Variables	B. Malgrange
14.	On Lie Groups and Representations of Locally Compact Groups	F. Bruhat
15.	On Mean Periodic Functions	J.P. Kahane
16.	On Approximation by Polynomials*	J.C. Burkill
17.	On Meromorphic Functions	W.K. Hayman
18.	On the Theory of Algebraic Functions of One Variable*	M. Deuring
19.	On Potential Theory*	M. Brelot
20.	On Fibre Bundles and Differential Geometry	J.L. Koszul
21.	On Topics in the Theory of Infinite Groups*	B.H. Neumann
22.	On Topics in Mean Periodic Functions and the Two-Radius Theorem	J. Delsarte
23.	On Advanced Analytic Number Theory[1]	C.L. Siegel
24.	On Stochastic Processes	K. Ito
25.	On Exterior Differential Systems	M. Kuranishi
26.	On Some Fixed Point Theorems of Functional Analysis*	F.F. Bonsall
27.	On Some Aspects of p-adic Analysis	F. Bruhat
28.	On Riemann Matrices	C.L. Siegel
29.	On Modular Functions of One Complex Variable	H. Maass
30.	On Unique Factorization Domains*	P. Samuel
31.	On the Fourteenth Problem of Hilbert	M. Nagata

Not available

Available as volume 9, T.I.F.R. Studies in Mathematics